Rod Bursell

D1765153

Computational
Category Theory

Prentice Hall International
Series in Computer Science

C. A. R. Hoare, Series Editor

BACKHOUSE, R. C., *Program Construction and Verification*
BACKHOUSE, R. C., *Syntax of Programming Languages: Theory and practice*
DE BAKKER, J. W., *Mathematical Theory of Program Correctness*
BIRD, R., and WADLER, P., *Introduction to Functional Programming*
BJÖRNER, D., and JONES, C. B., *Formal Specification and Software Development*
BORNAT, R., *Programming from First Principles*
BUSTARD, D., ELDER, J., and WELSH, J., *Concurrent Program Structures*
CLARK, K. L., and MCCABE, F. G., *micro-Prolog: Programming in logic*
DROMEY, R. G., *How to Solve it by Computer*
DUNCAN, F., *Microprocessor Programming and Software Development*
ELDER, J., *Construction of Data Processing Software*
GOLDSCHLAGER, L., and LISTER, A., *Computer Science: A modern introduction (2nd edn)*
GORDON, M. J. C., *Programming Language Theory and its Implementation*
HAYES, I. (ED.), *Specification Case Studies*
HEHNER, E. C. R., *The Logic of Programming*
HENDERSON, P., *Functional Programming: Application and implementation*
HOARE, C. A. R., *Communicating Sequential Processes*
HOARE, C.A.R., and SHEPHERDSON, J. C. (EDS), *Mathematical Logic and Programming Languages*
HUGHES, J. G., *Database Technology: A software engineering approach*
INMOS LTD. *occam Programming Manual*
INMOS LTD. *occam 2 Reference Manual*
JACKSON, M. A., *System Development*
JOHNSTON, H., *Learning to Program*
JONES, C. B., *Systematic Software Development using VDM*
JONES, G., *Programming in occam*
JOSEPH, M., PRASAD, V. R., and NATARAJAN, N., *A Multiprocessor Operating System*
LEW, A., *Computer Science: A mathematical introduction*
MacCALLUM, I., *Pascal for the Apple*
MacCALLUM, I., *UCSD Pascal for the IBM PC*
MEYER, B., *Object-oriented Software Construction*
PEYTON JONES, S. L., *The Implementation of Functional Programming Languages*
POMBERGER, G., *Software Engineering and Modula-2*
REYNOLDS, J. C., *The Craft of Programming*
RYDEHEARD, D. E., and BURSTALL, R. M., *Computational Category Theory*
SLOMAN, M., and KRAMER, J., *Distributed Systems and Computer Networks*
TENNENT, R. D., *Principles of Programming Languages*
WATT, D. A., WICHMANN, B. A., and FINDLAY, W., *ADA: Language and methodology*
WELSH, J., and ELDER, J., *Introduction to Modula-2*
WELSH, J., and ELDER, J., *Introduction to Pascal (3rd edn)*
WELSH, J., ELDER, J., and BUSTARD, D., *Sequential Program Structures*
WELSH, J., and HAY, A., *A Model Implementation of Standard Pascal*
WELSH, J., and MCKEAG, M., *Structured System Programming*
WIKSTRÖM, Å., *Functional Programming using Standard ML*

Computational Category Theory

DAVID E. RYDEHEARD
University of Manchester

ROD M. BURSTALL
University of Edinburgh

PRENTICE HALL

NEW YORK   LONDON   TORONTO   SYDNEY   TOKYO

First published 1988 by
Prentice Hall International (UK) Ltd,
66 Wood Lane End, Hemel Hempstead,
Hertfordshire, HP2 4RG
A division of
Simon & Schuster International Group

© 1988 D. E. Rydeheard and R. M. Burstall

All rights reserved. No part of this publication may be
reproduced, stored in a retrieval system, or transmitted, in any
form, or by any means, electronic, mechanical, photocopying,
recording or otherwise, without the prior permission, in
writing, from the publisher.
For permission within the United States of America contact
Prentice Hall Inc., Englewood Cliffs, NJ 07632.

Printed and bound in Great Britain by
A. Wheaton & Co. Ltd, Exeter.

Library of Congress Cataloging-in-Publication Data

Rydeheard, D. E. (David E.)
  Computational category theory.

  (Prentice Hall international series
  in computer science)
  Bibliography: p.   Includes index.
  1. Categories (Mathematics) – Data processing.
  2. Electronic digital computers – Programming.
  I. Burstall, R. M.   II. Title.   III. Series.
QA169.R93   1988   511.3   88-4232
ISBN 0-13-162736-8

British Library Cataloguing in Publication Data

Rydeheard, D. E., *1954–*
  Computational category theory.
  1. Computer systems. Applications of category
  theory
  I. Title   II. Burstall, R. M., *1934–*
  004'.01'51255
  ISBN 0-13-162736-8

---

1 2 3 4 5   92 91 90 89 88

ISBN 0-13-162736-8

Contents

Foreword
John W. Gray

Why should there be a book with such a strange title as this one? Isn't category theory supposed to be a subject in which mathematical structures are analyzed on such a high level of generality that computations are neither desirable nor possible? Historically, category theory arose in algebraic topology as a way to explain in what sense the passages from geometry to algebra in that field are 'natural' in the sense of reflecting underlying geometric reality rather than particular representations in that reality. The success of this endeavor led to many similar studies of geometric and algebraic interrelationships in other parts of mathematics until, at present, there is a large body of work in category theory ranging from purely categorical studies to applications of categorical principles in almost every field of mathematics. This work has usually been presented in a form that emphasizes its conceptual aspects, so that category theory has come to be viewed as a theory whose purpose is to provide a certain kind of conceptual clarity.

What can all of this have to do with computation? The fact of the matter is that category theory is an intensely computational subject, as all its practitioners well know. Categories themselves are the models of an essentially algebraic theory and nearly all the derived concepts are finitary and algorithmic in nature. One of the main virtues of this book is the unrelenting way in which it proceeds from algorithm to algorithm until all of elementary category theory is laid out in precise *computational* form. This of course cannot be the whole story because there are some deep and important results in category theory that are non-constructive and that cannot therefore be captured by any algorithm. However, for many purposes, the constructive aspects are central to the whole subject.

This is important for several reasons. First of all, one of the most important features of category theory is that it is a *guide to computation*. The conceptual clarity gained from a categorical understanding of some particular circumstance in mathematics enables one to see how a computation of relevant entities can be carried out for special cases. When the special case is itself very complex, as frequently is the case, then it is a tremendous advantage to know exactly what one is trying to do and in principle how to carry out the computation. The idea of mechanizing such computations is very intriguing. The present book, of course, does not enable one to do this, but it can be viewed as an essential precursor of developments that will lead to such mechanization. Categories themselves must be present in the computer as well as many particular examples of them before mechanical computation of categorical entities can be carried out.

Secondly, the fact that category theory is essentially algebraic means that it can be learned by learning these basic constructions. It comes as something of a shock to realize that one aspect of category theory is that it is 'just' a collection of ML-algorithms. However, it is particularly important for computer scientists and students of computer science that there is such a programming language representation of the subject. Because mathematicians have accumulated geometric and algebraic intuitions, many things can be elided in presenting category theory to them. But computer scientists generally lack these intuitions, so these elisions can present a great difficulty for them. Computer code does not permit such elisions and thus presents the basic material in a form that reassures computer scientists and allows them to use their intuitions for and understanding of programs to gain an advantage similar to the mathematicians' advantage from their knowledge of geometry and algebra.

Of course, all of this is beside the point unless there is a reason for computer scientists to need to learn category theory. However, the reasons are easily found by looking into almost any issue of a journal in theoretical computer science. Either the category theory is explicitly there or should be there and is missing only at the expense of devious circumlocutions. It really cannot be avoided in discussing the semantics of programming languages. The most dramatic instance of this arises in the semantics of the polymorphic lambda calculus which underlies ML. It really is an engaging thought that one needs category theory to explain ML, while in turn ML is a vehicle for explaining category theory.

That brings up the last point. There is another audience for this book; namely, category theorists who want to understand theoretical computer science so that they can participate in the exciting interactions that are taking place between these two fields. One very important entry point into the problems of theoretical computer science is just to examine computer programs and to wonder what they mean. There probably is no final answer to this question, but along the way, this book can serve as an invaluable stimulus to further research.

Preface

This is an account of a project we have undertaken in which basic constructions of category theory are expressed as computer programs. The programs are written in a functional programming language, called ML, and have been executed on examples. We have used these programs to develop algorithms for the unification of terms and to implement a categorical semantics.

This book should be helpful to computer scientists wishing to understand the computational significance of theorems in category theory and the constructions carried out in their proofs. Specialists in programming languages should be interested in the use of a functional programming language in this novel domain of application, particularly in the way in which the structure of programs is inherited from that of the mathematics. It should also be of interest to mathematicians familiar with category theory – they may not be aware of the computational significance of the constructions arising in categorical proofs.

In general, we are engaged in a bridge-building exercise between category theory and computer programming. Our efforts are a first attempt at connecting the abstract mathematics with concrete programs, whereas others have applied categorical ideas to the *theory* of computation.

The original motivation for embarking on the exercise of programming categorical constructions was a desire to get a better grip on categorical ideas, making use of a programmer's intuition. The abstractness of category theory makes it difficult for many computer scientists to master it; writing code seemed a good way to bring it down to earth. Someone with a computing background who wishes to learn category theory should have recourse to standard texts, some of which are listed later, but could well find this book a helpful companion text. Mathematicians who have learned a little programming, perhaps in conventional languages like Pascal, may profit from seeing how the functional programming style can embody abstract mathematics and do it in a way not too far from mathematical notation.

In preparing this book, we would especially like to thank John Gray for contributing a foreword. His enthusiasm for this project will be evident. Tony Hoare and the referees gave detailed comments for improving the book. Mike Spivey carefully read the manuscript and gave some useful comments. Anne Rydeheard and John Stell undertook some proofreading for which we are grateful. Ma Qing Ming and Don Sannella pointed out some errors in an early draft. Finally, we are indebted to LaTeX2 and Microsoft Word 3, two document preparation systems used for the book.

Chapter 1

Introduction

The usual occupation in computer science is to build a tool of some kind, for example, a compiler or a window manipulation package. In pure mathematics, on the other hand, we define new entities, for example, complex numbers, and demonstrate their properties. The motivation in this book is neither of these; rather it is to illustrate a connection between two hitherto widely separated branches of knowledge: computer programming and category theory. We started off in a spirit of creative play, programming some basic constructions in category theory. We hoped that it would provide a tool for advanced programming, harnessing the abstraction of category theory for use in program design. These hopes have not really been realized by our work so far. However, we do present two example applications where categorical constructions are used in the development of programs. More immediately our work has educational value for both computer scientists and mathematicians.

Category theory should have a particular interest for computer scientists because it seems to operate on the same level of generality as logic and computer programming. None of these are committed to any particular branch of mathematics, such as algebra or number theory. The essential virtue of category theory is as a discipline for making definitions, and making definitions is the programmer's main task in life. What else is the programmer doing when she writes code? Somehow categorical definitions come in larger chunks than definitions of individual functions in a program. Notably, when we define the adjoint to a functor, we get a new functor (a parameterized data type), a natural transformation (a function) and a bijection between hom-sets (another function). Thus an adjoint definition corresponds to a module in a programming language rather than a single function definition, but it is a module which has

1

some internal cohesion and *raison d'être*, instead of a bundle of functions which the modularly-minded programmer has forced into uneasy proximity.

Another reason why computer scientists might be interested in category theory is that it is largely constructive. Theorems asserting the existence of objects are proven by explicit construction. This means that we can view category theory as a collection of algorithms. These algorithms have a generality beyond that normally encountered in programming in that they are parameterized over an arbitrary category and so can be specialized to different data structures.

We have expressed categorical algorithms in ML, a functional programming language. Functional languages are closer to mathematical notation than are imperative languages like Basic or Pascal. One writes expressions to denote mathematical entities rather than defining the transitions of an abstract machine. ML also provides types which make a program much more intelligible and prevent some programming mistakes. ML has polymorphic types which allow us to express in programs something of the generality of category theory. However, the type system of ML is not sufficiently sophisticated to prevent the illegal composition of two arrows whose respective source and target do not match. This requires a computation of equality on objects. It is an open question whether a programming language with dependent types or a subtype mechanism can do better.

The relationship of the mathematics to the ML code is as follows: (1) categorical concepts are represented as types in ML, and (2) constructive proofs of theorems in category theory become ML programs. For instance, the theorem that if a category has an initial object and pushouts then it has all finite colimits yields an iterative algorithm for constructing the colimiting cocone of a finite diagram, starting with the initial object and using the pushout at each iteration.

We should make it clear that we have not invented a new programming language or a new specification language. We simply used an existing functional language, ML, to write a novel kind of program of unusual generality. Tatsuya Hagino has indeed invented such a new language for programming and specification, based on adjoints. It turns out to be very like ML, almost identical in its expressive power, but using fewer primitive notions and hence having a more rational structure, a sort of natural mathematical unfolding of the main language concepts as opposed to a computer science evolution of them by trial and error of language designers. We say a little about Hagino's work in Chapter 10.

It has been clear for a long time that the many of the proofs in category theory are constructive and hence could be translated into algorithms; so in a mathematical sense we have just spelled out the obvious. However, from a programming point of view, there is considerable interest in seeing carefully worked out programs to represent the essence of the categorical proofs and to notice that these programs have a certain elegance and pleasing structure. We went to considerable trouble through various formulations to embody as much of the elegance of the categorical approach as possible in our programs. For example, having written a certain function which we needed, we noticed that it formed the object part of a functor and that the arrow would be helpful later on. Seeing these two functions as part of the same functor is a good example of categorical thinking imposing mathematical structure on a program. The Nuprl system [Constable et al. 85] is a proof development system based on constructive logic which automatically extracts a program from a proof. It would be interesting to see how such automatically generated programs compare with our hand-coded ones. Probably in Nuprl one could obtain elegant programs by creating a proper organization of the proof, but the question is as yet unexplored. Unlike the Nuprl formulation, our algorithms only represent part of the information in a proof; they embody the construction; the remaining information in the proof corresponds to the verification showing that the construction produces the required result.

In programming category theory, we are confronted at the outset by the problem: how do we represent a category? Do we use a list of objects and a list of arrows? This would mean we represent only finite categories. Instead we use a functional representation in which the class of objects and that of arrows are types in ML. This allows us to represent infinite categories. Another representation problem arises with the ubiquitous universal properties of category theory. Again we make use of functions, in this case higher order functions. The programs derived from categorical constructions are parameterized on categories. In order to apply the programs to a range of categories, we need systematic ways of constructing categories rather than explicitly encoding them. Goguen suggested we use comma categories for computations on structures such as graphs. We have also made use of functor categories. Another aspect of category theory that is used in the programming is duality. Duality is a fundamental principle in category theory arising from the invariance of the theory under the reversal of arrows. We use it, for instance, to convert programs computing colimits to those computing limits.

In a final chapter we discuss other approaches to computational representation of category theory, notably those of Dyckhoff and Goguen, which are similar in spirit to ours, and that of Hagino, which differs rather radically and interestingly.

We have discovered that applications of our categorical approach to specific computing problems are not easily developed. You have to really understand a task to abstract it in a categorical framework. However, we have two quite interesting applications, a general unification algorithm using coequalizers, which specializes to known unification algorithms, and a categorical implementation of the specification constructing operations in the language Clear.

Since the early 1970s there has been an increasing amount of interest in using category theory to explicate aspects of the theory of computation, in particular, the semantics of programming languages. This is somewhat outside the scope of this book although we try to indicate where categorical concepts are relevant to programming. The range of applications of category theory in computation may be judged from the proceedings of two conferences published as Lecture Notes in Computer Science, nos. 240 (1986) and 283 (1987), Springer-Verlag.

1.1   The contents

In the succeeding chapters, we describe the techniques used in the programming of category theory.

In Chapter 2, we describe the functional programming language Standard ML. We cover all the features of ML that we use later in the book, using illustrative examples. This is meant as a tutorial and a series of exercises is included. Answers to these exercises may be found in an appendix to the book. Those with knowledge of ML can safely omit this chapter. Those with some experience of functional languages may wish to browse through the chapter to acquaint themselves with the syntax of ML. Others ought to read the chapter so as to be able to understand the subsequent programming. In Appendix A there is an index of ML keywords. This may be used as a reference for reading ML programs. Programming in ML is often a rewarding experience and we encourage the reader to get hold of an ML system to practice on.

Chapters 3 to 7 lay out basic category theory. We describe the mathematical concepts and constructions and the corresponding ML programs. We choose illustrative examples which are relevant to programming rather than those drawn from areas of abstract mathematics. In this, we hope to avoid relying on mathematical intuition but instead, through the programs, use programmer's insight to get over the abstract concepts of category theory.

In Chapter 3, we present categories and functors. We define them and show how to represent them in ML. Illustrative examples are included. We also consider the principle of duality, coding it as operations on categories and functors. In Chapter 4, we describe how universally defined concepts may be represented. We deal with limits and colimits. We also present the first substantial programs which arise from categorical constructions of colimits. Duality allows us to convert these into programs computing limits.

Chapter 5 introduces constructions of categories. We concentrate on comma categories and functor categories. In each case, under certain conditions, colimits in constructed categories may be computed from those in the component categories. We program up this inheritance. By introducing canonical isomorphisms, duality allows us to convert this inheritance of colimits to an inheritance of limits.

In Chapter 6, we look at adjunctions. Adjunctions occur widely in mathematics and programming. We define them and represent them as an ML type. We introduce constructions of adjunctions based on the 'term algebra' construction of free algebras. In Chapter 7, we briefly introduce toposes taking the theory as far as internal logics within toposes. We display programs to compute internal logics.

Chapters 8 and 9 are applications of the categorical programming. In Chapter 8, we consider the unification of terms. This is a task arising in the automation of inference and corresponds to solving equations. We show how unification algorithms may be derived from constructions of colimits. In Chapter 9, we look at colimits in a different role: the construction of algebraic theories. We implement the semantics of the algebraic specification language, Clear. Operations for combining theories are described in terms of colimits in certain categories.

Finally, in Chapter 10, we discuss formal (linguistic) aspects of category theory. We list some requirements on a formalism for expressing category theory. We also look at fragments of category theory in formalisms other than ML. We present an algebraic treatment in OBJ due to Goguen and a description of category theory in a constructive type theory due to Dyckhoff. We also briefly describe an interesting system of Hagino, which consists of a programming language based upon the universal concepts of category theory.

We include most of the programs we have written so that this book may serve as a manual to those wishing to use the categorical programming. At the back of the book in Appendices B and C there is an index of the functions that we have defined. This provides a cross-reference for reading the categorical programs.

Some sections in the book are starred. These may safely be omitted as they contain material somewhat aside from main development.

Exercises will be found scattered throughout the book, mainly at the end of chapters. Some of these are meant to reinforce the reader's understanding or introduce further examples of what has already been covered. Some, however, are starred exercises. These are more substantial and explore new topics in the form of mini-projects or open-ended questions.

1.2   Accompanying texts

In this section we give details of some books which the reader may find useful to complement the material of this book.

1.2.1   Textbooks on category theory

This book does not have the breadth or depth of coverage of a mathematical text on category theory. Here topics are chosen for their computational significance. Categorical texts are aimed at a mathematical audience and some require a fairly substantial mathematical background.

We list some textbooks on category theory which expand upon the material in this book:

Arbib, M. and Manes, E. (1975) *Arrows, Structures and Functors: The Categorical Imperative.* Academic Press, London.

Barr, M. and Wells, C. (1985) *Toposes, Triples and Theories.* Grundlehren der mathematischen Wissenschaften, 273, Springer-Verlag, New York.

Goldblatt, R. (1979) *Topoi – The Categorial Analysis of Logic.* Studies in Logic and the Foundations of Mathematics, 98, North-Holland, Amsterdam.

Herrlich, H. and Strecker, G.E. (1973) *Category Theory.* Allyn and Bacon.

Lambek, J. and Scott, P.J. (1986) *Introduction to Higher-order Categorical Logic.* CUP.

Mac Lane, S. (1971) *Categories for the Working Mathematician.* Springer-Verlag, New York.

Schubert, H. (1972) *Categories.* Springer-Verlag, Berlin.

1.2.2 ML references and availability

The two references below are full and readable reports on the version of the ML language used in this book, called Standard ML:

Harper, R. and Mitchell, K. (1986) *Introduction to Standard ML.* Unpublished report, University of Edinburgh.

Wikström, A. (1987) *Functional Programming using Standard ML.* Prentice Hall International, Hemel Hempstead.

The Standard ML system runs on most machines that support UNIX (©Bell Laboratories). It is available from:

Laboratory for Foundations of Computer Science,
Department of Computer Science,
University of Edinburgh,
The King's Buildings,
Edinburgh EH9 3JZ

1.2.3 A selection of textbooks on functional programming

Abelson, H. and Sussman, G.J with Sussman, J. (1985) *Structure and Interpretation of Computer Programs.* MIT Press, Cambridge, Mass.

Darlington, J., Henderson, P. and Turner, D.A. (1982) *Functional Programming and its Applications: An Advanced Course.* CUP.

Glaser, H., Hankin, C. and Till, D. (1984) *Principles of Functional Programming.* Prentice Hall International, Hemel Hempstead.

Henderson, P. (1980) *Functional Programming: Applications and Implementation.* Prentice Hall International, Hemel Hempstead.

1.3   Acknowledgements

We have spent a few years now on this project and over this time many
people have given us help and support.

We are indebted to Don Sannella at Edinburgh who was a collabo-
rator on this project. Don undertook some of the early programming in
a hectic few weeks one summer. Later he jointly undertook the work
described in Chapter 9. Throughout, Don has helped with his expertise
on functional programming. Here at Manchester, John Stell spent a few
months translating the code into Standard ML.

Amongst others who have helped, Joseph Goguen deserves special
mention. He collaborated with one of the authors (RMB) in developing a
categorical model theory of program specification. It was this work that
prompted RMB to consider computational aspects of category theory,
see [Burstall 80]. Throughout, he has kept an encouraging interest in
the project and given useful guidance, including turning our attention to
some categorical algorithms involving colimits and operations on graphs
as well as encoding a fragment of category theory in a version of OBJ.

We are especially grateful for the Foreword provided by John W.
Gray. His enthusiasm for this project will be evident.

David MacQueen used this categorical programming to test ideas
about modularity in programming. Roy Dyckhoff kindly allowed us to
describe his implementation of category theory and gave detailed comments on the account in this book. Both Horst Reichel and Ursula Martin
helped in the development of the unification algorithm. Esther Dennis-
Jones helped with modular programs in ML and Victoria Stavridou with
OBJ. We are indebted to all those who, under the leadership of Robin
Milner, developed the programming language Standard ML.

Others who have contributed through discussions, advice and en-
couragement are Gordon Plotkin, David Benson, Andrzej Tarlecki, Peter
Aczel and Michael Barr.

The work described in this book was undertaken with financial support from the Science and Engineering Research Council.

Chapter 2

Functional Programming in ML

This chapter is an introduction to some aspects of functional programming. It is also a guide to the programming language ML. In the next few chapters we shall use ML to program constructions from category theory. Those familiar with the ML language can safely omit the chapter. An index of ML keywords appears at the end of the book for reference when reading ML programs.

Functional programming arose from two sources: the execution of programs on computing machinery, and logicians' interest in languages and functions. Early computer programs consisted of sequences of machine instructions. It was found difficult to reason about these programs so as to establish their behaviour. Part of the problem was the linguistic distance between the primitive instructions and the intended behaviour. Moreover, the simple substitution properties normally associated with variables were not applicable to these programs. The action associated with an instruction depended not only on the context of the instruction in the program but also upon the state of the machine, which is determined by the computational history of the program. Landin [1966] pointed out that many of the features of programming languages could be maintained whilst at the same time having the usual substitution rules. McCarthy [1960] had previously developed such a programming language called Lisp. Many other 'functional' programming languages have been developed since then. These are based on defining functions and satisfy, in the main, the standard substitution rules.

9

The other historical strand is the development by logicians of languages for describing functions. This was an attempt to treat functions as primitive, defined by 'rules', rather than as graphs defined in terms of sets. Two directions were pursued. Combinatory logic [Curry, Feys 68] provides a variable-free description of functions. Backus [1978] has proposed that variable-free languages, combinator languages, be developed for programming. At approximately the same time as combinatory logic was being developed, Church [1941] proposed the λ-notation for functions and associated calculi. λ-calculus has been influential in the theory and design of modern programming languages.

Our interest is in languages for describing mathematical constructions. Constructions transform input to output, taking structures of one kind to structures of another. This dependence of output on input is functional and makes a functional programming language a suitable vehicle for programming constructions like those in category theory. We describe here the functional programming language ML, which we have used to program category theory. ML was developed by Milner [1978] as a language for constructing mathematical proofs on a computer – hence the name: it is a metalanguage (ML) for proof development. ML owes a good deal to predecessors such as Lisp [McCarthy 60] and ISWIM [Landin 66] as well as to λ-calculus. It resembles other programming languages like Miranda [Turner 86] and OBJ [Goguen, Tardo 79]. The novelty is its type system, with polymorphic types and a type definition mechanism. The language is based upon higher-order recursion equations for function definition. It incorporates a type abstraction mechanism and support for modular programming [MacQueen 85]. Various non-functional features, like exceptions and references, have been found convenient and been incorporated into the language. ML can be used interactively. Types and functions may be defined and expressions evaluated in a fairly simple manner. Readers are encouraged to try programming in ML; they should find it a pleasant experience.

The version of ML that we describe, known as Standard ML, was developed by a team at Edinburgh under the leadership of Robin Milner [1984]. It incorporates features of previous versions of ML as well as from the language Hope [Burstall, MacQueen, Sannella 80]. Full and readable reports of the language are [Wikström 87] and [Harper, Mitchell 86]. The latter has proved useful in compiling this chapter. In the following description we concentrate on those features of ML that we use in the programming of category theory.

2.1   Expressions, values and environments

The evaluation of expressions, by which syntactic descriptions are transformed into the values, is fundamental to programming.

An example expression for an integer in ML is (3+4)*5, where * denotes multiplication. All phrases to be evaluated in ML are terminated with a semicolon. Typing this expression followed by a semicolon will give the response,

```
> 35 : int
```

which consists of the value 35 and its associated type int, that of integers.

An example expression in truth-values (booleans), rather than integers, is:

```
not(true andalso false)
```

It will give as result:

```
> true : bool
```

In programming, we want not only to describe values, but also to name them for future reference. A collection of named values is called an *environment*. The syntax which evaluates to an environment is called a *(value) binding*.

A simple binding consists of an identifier (or 'variable', we shall use the terms interchangeably) m and its associated value described by an expression. For example:

```
val m = (3+4)*5
```

The response to this, the environment denoted by this binding, is obtained simply by evaluating the expression:

```
> val m = 35 : int
```

Here is another example binding, this time for truth-values:

```
val x = not(true andalso false)
```

It should be noted that bindings are *not* 'assignment' statements, like x := 35 in Pascal. A binding gives a value to a variable just once at the point where the variable is declared. The value of an expression in the presence of bindings depends only on its textual context and not on some notion of computational history. This is called *referential transparency*

and distinguishes functional languages from imperative languages which
have assignment and updating operations. Standard substitution rules
are valid in functional programming, making for easier reasoning about
and manipulation of programs.

Larger environments can be built by combining bindings. There is
a *parallel* combination of bindings in which each binding is evaluated
independently. To avoid clashes of definition the variables bound should
be distinct. An example of a parallel combination is:

```
val m = (3+4)*5 and n = 6*7
```

This results in the following environment:

```
> val n = 42 : int
  val m = 35 : int
```

Bindings may also be combined *sequentially* in which case the expression in the second can use the variable bound in the first, as illustrated
below.

```
val m = (3+4)*5; val n = 6*m
```

This evaluates to the following environment:

```
> val m = 35 : int
> val n = 210 : int
```

An expression containing variables can be evaluated by supplying
values to the variables, that is, it can be evaluated in the presence of a
suitable environment. This is described using a `let`-clause:

```
let val m = (3+4)*5 in m*m + (m+1)*(m+1) end
```

The result of this is simply the value of the expression with the variables
instantiated to their values given in the binding:

```
> 2521 : int
```

Notice that the binding of m to its value is not available after the expression is evaluated. This is described by saying that the variable and its
binding is *local* to the clause.

Conditional expressions are available in ML and are written in the
familiar form:

```
if x=1 then 0 else 2+x
```

2.2 Functions

Functional programming is about defining, naming and invoking functions. A function f is applied to a value v (its argument) by juxtaposition f v. Parentheses may be inserted if wished to give the standard notation f(v).

This is for *prefix* functions – those in which application is by prefixing the function name to the argument. Other syntax for application is in use. For instance, binary operations are often *infix*, like the addition of numbers $3 + 4$. To introduce infix operations we declare their infix nature together with a precedence – a number used to disambiguate expressions; the higher the number the tighter it binds its arguments:

```
infix 4 +
```

Functions may be defined using the standard mathematical format:

```
fun f(x) = 2*x
```

The result of typing this into ML is:

```
> val f = fn : int -> int
```

This tells us that we have defined an environment in which the variable f denotes a function which cannot be printed. The type information consists of two type expressions separated by an arrow; the first type is that of the argument to the function, the second is the type of the result of evaluating the function.

For numerical values, we can separate the definition of a value from its name. To do this for functions we introduce expressions denoting functions. Here is an example.

```
fn x => 2*x
```

This denotes the function which doubles an integer, so we may write,

```
(fn x => 2*x)(3)
```

to apply the function to the value 3 to yield the result 6. Notice how the result is obtained. The variable x is bound to the argument 3 and the expression 2*x is then evaluated. This notation is a variant of the λ-expression $\lambda x.(2 \times x)$. Notice that

```
fun f(x) = 2*x
```

is equivalent to

```
val f = fn x => 2*x
```

ML is *statically scoped*. Free variables in expressions are resolved in the context of the definition of the expression rather than the context of its evaluation.

2.2.1   Recursive definitions

Function definitions may be recursive, in which case the function being defined occurs in the expression on the right-hand side of the definition. The factorial function, $n! = 1 \times 2 \times \ldots \times n$, can be defined recursively as:

```
fun factorial(x) = if x = 0 then 1 else x*factorial(x-1)
```

An application of the factorial function `factorial(n)` is evaluated by a successive 'unfolding' of the definition until the base case `factorial(0)` is reached.

Here is a recursive definition of a function which tests whether an integer is even or not:

```
fun even(x) = if x=0 then true else
                 if x > 0 then not(even(x-1))
                          else not(even(x+1))
```

Recursion is the key to defining a wide range of functions in a functional programming language like ML. As a repetitive construct it replaces iteration in imperative languages (like the WHILE loop in Pascal). The relationship between recursion and iteration, the uses and efficiency of each, is a fairly involved topic for which the reader should consult a reference (e.g. [Kruse 87]). Recursion introduces the possibility of non-termination in programs. Termination proofs are needed to ensure the well-formedness of definitions.

2.2.2   Higher order functions

Higher order functions are functions that take functions as arguments or return them as results. These are available in ML. This is an important aspect of programming as many mathematical structures, such as automata, algebras, categories and adjunctions, are functional in nature and therefore constructions on them are inherently higher order.

Let us consider some examples.

A simple and rather useless example of a function taking a function as argument is the function `eval_at_one` which takes a function f acting on integers and returns the value `f(1)`:

```
fun eval_at_one(f) = f(1)
```

Thus `eval_at_one(factorial)` has value 1 (= 1!) and

```
eval_at_one(fn x => if x > 0 then true else false)
```

has value `true`.

Here is a more interesting example,

```
fun poly_eval(f) = f(f(3)) + f(3) + 3
```

which on argument `factorial` will result in the value 729 (= 6! + 3! + 3).

Functions may also be returned as results:

```
fun add_on(m) = fn n => m + n
```

Thus `add_on(3)` is the function which adds 3 to a number. Definitions like these go under the name of *partial evaluation* as some of the arguments are supplied and some are left uninstantiated to form a function. We may write this function definition equivalently using multiple arguments:

```
fun add_on(m)(n) = m + n
```

As a final example we give a function which has functional arguments and functional results:

```
fun poly(f)(x) = f(f(x)) + f(x) + x
```

Thus `poly_eval(f) = poly(f)(3)`.

2.3 Types

Types are introduced into programming languages to organize the space of values by dividing it into sections, each section being identified with a type. The point of this is to gather together values of the same form so that the validity of applying functions may be controlled. Where there is a range of different values, definitions of functions presuppose a certain form for the argument. For example, integer addition is a different algorithm from real addition and, moreover, the algorithm only applies to values which have the form of an integer so cannot be applied to either real numbers or truth-values.

We write v:T to denote that value v is of type T. In ML all values have an associated type (such languages are called *strongly-typed*). Moreover, the fact that a value has a particular type is recognizable from the form of the expression for the value. Resolving the type of expressions is called type checking and languages like ML are designed so that type checking is decidable. Whilst this restricts the expressiveness of the type system, it allows for checking the type well-formedness of programs before they are run, thus eliminating one source of errors in programming.

2.3.1   Primitive types

The primitive types in ML are integers, real numbers, truth-values (also called booleans), strings (of characters) and the unit type.

The numerical types, integers and reals, have the usual arithmetic operations defined upon them as well as equality and the various inequalities. These operations are overloaded in the sense that the same symbol is used for operations on integers and on reals. This means that occasionally we have to disambiguate expressions by explicitly including type information. For example:

```
fun add(x:int,y:int) = x+y
```

The type of truth-values is called bool and has values true and false. The operations 'not' not, 'and' andalso, and 'or' orelse are available.

Strings are finite sequences of characters and are written inside double quotes e.g. "the quick brown fox". Strings may be concatenated end-to-end (using infix ^). The operation size returns the length of a string.

Finally, the unit type unit consists of a single value, written (). It is a formal device used, for example, to make constants into functions with no arguments.

2.3.2   Compound types

We have already seen a type building operation – from types A and B we may form the type A -> B of all functions from A to B.

The type of *tuples* consists of sequences of values of a fixed length enclosed in parentheses. If a:A, b:B, ... d:D then the tuple (a,b,...,d) is of type A*B*...*D, sometimes called the product type. For instance:

```
(2,true,"brown") : int * bool * string
```

To name the components in a tuple, record types are available. The following is a record type for personal files:

```
{name: string, salary: int, gender: bool}
```

Values of this type are given as:

```
{name="fred", salary=10000, gender=true}
```

A further type formation operation, that of lists, is available and discussed in Section 2.6.

2.3.3 Type abbreviation

Names may be given to type expressions:

```
type Fcn_and_Int = (int -> int) * int
```

This is not the creation of a new type. Equality of types is structural equality and type names get expanded to their definitions.

2.4 Type polymorphism

In languages where values have associated types, it seems natural to associate just one type with each value. Such languages are called monomorphic. This, however, is unnecessarily restrictive as there are constructions which are uniform over a range of types. A single program can act on values of various types. To gain this generality, a type system must allow values to have more than one type. Languages with such type systems are called polymorphic.

Type polymorphism arose in combinatory logic [Hindley 69] and λ-calculus [Girard 72] and was introduced into programming by Strachey [1967], Reynolds [1974] and by Milner [1978] in the language ML. Strachey makes the distinction between polymorphism based upon a uniformity of action, which he calls 'parametric' polymorphism and polymorphism based upon a common name for an operation, which he calls *ad hoc* polymorphism. This 'overloading' of the name of an operation is common and useful, for instance in arithmetic + stands for the addition of integers and of real numbers. The programs for these two operations may be quite different. Again, a print function takes values and transform them into strings of characters for display. A different transformation is required for different types of the argument.

Parametric polymorphism with its uniformity of action is a more fundamental notion which we illustrate here with some examples.

Consider the simple function which projects a pair onto the first argument, defined as follows:

```
fun first(x,y) = x
```

It has many different types. If x and y are integers, then it has type int*int -> int, so that first(2,4) is well-typed. If x is an integer and y is a truth-value, then the function has type int*bool -> int and first(2,true) is well-typed.

By introducing *type variables*, we can give an expression which encompasses all the types of the function first. The *most general type* of the function first is:

```
first : 'a * 'b -> 'a
```

Type variables are distinguished in ML (for parsing purposes) by an initial quotation mark. Any type for the function is obtained by instantiating the type variables.

The type system of ML is such that all well-formed expressions have a unique most general type and this can be determined from the expression using a unification algorithm.

As another example of type polymorphism, consider the higher order function:

```
fun twice(f) = fn x => f(f(x))
```

Thus twice(square)(3) is square(square(3)), and so is 81. The most general type of this function is given by the expression:

```
twice : ('a -> 'a) -> ('a -> 'a)
```

Notice that, in the definition of twice, the argument type of f must be the same as its result type, since f is applied to itself on the right-hand side.

Structures which store values often admit operations which change the structure independently of the type of the values stored. Simple examples of this polymorphism are provided by list processing functions. Lists are linear sequences of items. They are homogeneous, meaning that all items within a list have have the same type. Thus [1,2,3] and [true,false,true,true] are lists, of numbers and truth-values respectively. [1,true,6] is not a list in this sense. Any list may be built from the empty list by successively putting items on the front of the list. Let nil denote the empty list and :: the operation which takes an item and a list and creates a new list consisting of the old list with the item on the front. Expressions built from nil and :: are in 1-1 correspondence with lists, e.g. 1::(2::(3::nil)) is the list [1,2,3].

Consider the function **append**, which concatenates two lists end to end:

append([1,3,2],[3,4]) = [1,3,2,3,4]

This clearly acts independently of the type of the items in the lists and so is polymorphic. It can be applied to any two lists, whatever the type of the items, as long as the two lists have the same type of items. What then is the most general type of this function? First, we need the type of homogeneous lists which we write as 'a list where 'a is a type variable. Thus the type of lists of integers is int list and that of reals is real list. Later, we shall see to define this type, at the moment we are concerned with types of list processing functions.

Using type variables to denote the polymorphic nature of functions we may write the (most general) type of the **append** function as:

append: ('a list)*('a list) -> ('a list)

By instantiating the type variable we see that **append** has many different types corresponding to the different types of lists to which it may be applied, for example,

append: (int list)*(int list) -> (int list)

as well as:

append: (bool list)*(bool list) -> (bool list)

Here are some further polymorphic functions on lists:

reverse : 'a list -> 'a list
nil : 'a list

Not all list processing functions are polymorphic. For instance, the function that adds up a list of integers is of type:

sum : int list -> int

2.5   Patterns

Components of structures are obtained by *pattern matching*, matching a *pattern* against a value. For instance, suppose that **v** is the triple (3,false,4):

```
val v = (3,false,4)
```

We may match this against a pattern (x,y,z) to create an environment in which x, y and z are bound to the corresponding components of v:

```
val (x,y,z) = v

> val z = 4 : int
  val y = false : bool
  val x = 3 : int
```

If only some of the components are required, the underscore can be used to prevent binding:

```
val (x,y,_) = v
```

The result of this will be the environment:

```
> val y = false : bool
  val x = 3 : int
```

Variables must not be repeated in a pattern. The following is incorrect as it would not be clear what type x is, let alone its value:

```
val (x,x,z) = v
```

A pattern, then, is an expression built from variables, value constructors (e.g. parentheses for constructing tuples) and the underscore, such that each variable in the expression occurs only once.

The simplest case of a pattern is a variable:

```
val x = 3
```

Values other than tuples form patterns. Here are records:

```
val r = {name = ("joe","smith"), age = 40}
```

The surname may be obtained by the match:

```
val {name = (_,surname), age = _} = r
> val surname = "smith" : string
```

It is useful to be able to match simultaneously a pattern and a subterm of a pattern against a value. To do this we introduce *layered patterns*. Consider the value:

```
val v = ((1,2),3)
```

We match this against a layered pattern (using the keyword **as**):

```
val (x as (_,y),_) = v
```

This yields the following environment:

```
> val y = 2 : int
  val x = (1,2) : int * int
```

Pattern matching can be avoided when, as well as 'constructor' operations for forming values, there are 'destructor' operations for extracting components of structures. If there is a fixed set of type constructors then we can introduce a set of destructor operations to accompany them. It is in the presence of an extensible type system, when new types and their values can be defined, that pattern matching comes into its own.

2.6 Defining types

ML allows the definition of new types and their associated values. The values are described in terms of operations for forming values, called *data constructors*. Values consists of expressions built out of data constructors. Here is a simple example in which there are three data constructors, each of which is a nullary (constant) operation:

```
datatype Colour = red | blue | green
```

This corresponds to an enumerated type having three values. The vertical bar | separates the different forms of the values. It may be read as 'or'. The type consists of the disjoint (labelled) union of the forms of the values indicated.

Functions are defined over this type by pattern matching against the different cases:

```
fun warm(red)   = true
  | warm(blue)  = false
  | warm(green) = false
```

Here the patterns are simply the constant data constructors enumerated in the type definition. In such case analysis, it is wise to ensure that the cases defined by patterns are both distinct and exhaustive so as to admit only single-valued functions which are total. In fact, the cases are accessed sequentially. We may make use of this to abbreviate definitions.

The definition above is equivalent to a case statement written in the following form:

```
fun warm(x) = case x of
        red => true | blue => false | green => false
```

Consider now a type definition with non-constant data constructors:

```
datatype Plant = flower of string*int*Colour |
                 foliage of string*int
```

Values of this type are of two forms. For flowering plants we give their name, height and colour of flowers, e.g. flower("rose",3,red). For foliage plants, we give their name and height, e.g. foliage("fern",2). Again, functions are defined by pattern matching:

```
fun height(flower(_,n,_) = n
  | height(foliage(_,n) = n
```

Type definitions may be recursive to describe values which are expressions of an arbitrary size. Peano's recursive definition of natural numbers becomes:

```
datatype Num = zero | succ of Num
```

This defines natural numbers using succ as the successor function adding 1 to a number. Thus zero stands for 0 and succ(succ(zero)) for 2.

Functions are again defined by case analysis but, in general, definitions will be recursive corresponding to the recursive structure of the values:

```
fun even(zero) = true
  | even(succ(n)) = not(even(n))
```

Addition of natural numbers may be defined as follows:

```
fun add(zero,n) = n
  | add(succ(m),n) = succ(add(m,n))
```

Types such as lists and trees, which act as storage structures for
values of arbitrary type, are defined by type parameterization. Consider
the simple example of a pair of values of the same type:

```
datatype 'a Pair = pair of ('a * 'a)
```

Here 'a is a type variable and `Pair` is a *type constructor*. By instantiating
the type variable various types can be obtained:

```
pair(3,4) : int Pair
pair(true,false) : bool Pair
```

Functions over parameterized types may be polymorphic:

```
fun first(pair(x,y)) = x
```

This function has type,

```
first : 'a Pair -> 'a
```

for any type 'a.

A recursive, parametric type is *list*,

```
datatype 'a list = nil | 'a :: ('a list)
```

defining linear lists of items of any type (all items in the list must have
the same type, i.e. the lists are *homogeneous*). Thus lists are either empty
nil or consist of an item v on the front of a list s, v::s. Thus the list
[2,3,4] corresponds to 2::(3::(4::nil)). Lists are built into ML with
the square-bracket notation. We saw some list processing functions in
Section 2.4. Here are some more examples:

```
fun length nil = 0
  | length (h :: t) = 1 + length t

fun member(e,nil) = false
  | member(e,(h::t)) = if e=h then true else member(e,t)
```

Equality of values is handled as a special function in ML. The primitive types all have a pre-defined equality function on them. Compound
types and user-defined types have an equality where this may be determined structurally. Roughly, this means that equality is pre-determined
on all non-functional types. Types involving functions do not in general
admit an extensional equality. Any intended equality on functional types

and any equality different from structural equality can be supplied by the user.

The ML type system is one among many that have been proposed for programming languages and for proof systems. Polymorphism makes it more expressive than, say, Pascal and yet type checking is decidable. The ML type system does not include dependent types (as in the programming language Pebble [Burstall, Lampson 84]), type quantification (see [Cardelli, Wegner 85]), subtypes (as in OBJ [Goguen, Tardo 79]) or type universes (as in [Martin-Löf 82]).

2.7   Abstract types

Programming languages provide a range of types and type formation operations. Part of the process of programming is the choice of types in the language to represent given structures. There will not, in general, be a unique way to represent a given structure. Moreover, rarely will it be possible to choose an exact representation. Usually the representation contains more structure than is necessary.

What is required of a representation is that certain operations may be defined and have a given behaviour. It is desirable to separate the representation from the use of the type. The operations mediate between the representation and its use. They provide an 'interface' to the representation. With this separation, we may change the representation and redefine the operations, without needing to change programs using the type. To make this effective, the representation should be inaccessible outside its definition. The operations on a type determine how much of the representation is available and so, where there is superfluous structure in a representation, this may be hidden. This separation of representation from use is called 'data abstraction' and is a fundamental idea in the organization of programs.

Let us have a look at an example of an abstract data type:

```
abstype Mixture = mix of int*int*int
    with val cement = mix(6,0,0)
         and sand   = mix(0,6,0)
         and gravel = mix(0,0,6)
         and mortar = mix(1,5,0)
         and infill = mix(1,2,3)
         fun compound(parts:int, mix(c,s,g),
                      parts':int,mix(c',s',g')) =
```

```
            let val p = parts + parts'
                val cp = (parts*c+parts'*c') div p
                and sp = (parts*s+parts'*s') div p
                and gp = (parts*g+parts'*g') div p
            in mix(cp,sp,gp) end
end
```

Here we define a type, representing it in terms of a triple of integers, and operations on the type. The representation is not available outside the definition, in particular the data constructor mix will not be defined outside the type definition. The only way to manipulate mixtures is through the operations provided, which act as an interface to the representation.

Finite sets provide another example of data abstraction. Some programming languages incorporate finite sets as an in-built type. In ML, we represent this type in terms of other types. We show how sets are represented by lists of their elements, using data abstraction to hide the order in which the elements are stored and the multiplicity of elements in a list.

```
abstype 'a Set = set of 'a list
   with val emptyset = set([])
        fun is_empty(set(s)) = length(s)=0
        fun singleton(x) = set([x])
        fun union(set(s),set(t)) = set(append(s,t))
        fun member(x,set(l)) = list_member(x,l)
        fun remove(x,set(l)) = set(list_remove(x,l))
        fun singleton_split(set(nil)) = raise empty_set
          | singleton_split(set(x::s)) =
                              (x,remove(x,set(s)))
        fun split(s) =
            let val (x,s') = singleton_split(s) in
            (singleton(x),s') end
end
```

For abstract types, pattern matching is no longer appropriate, so we introduce functions to take values apart, called 'destructor functions'. An example in the type above is the function singleton_split. New functions on sets may be defined using these destructor functions:

```
fun cardinality(s) = if is_empty(s) then 0 else
            let val (x,s') = singleton_split(s) in
            1 + cardinality(s') end
```

We can avoid destructor functions and this explicit recursion if we can find a collection of operations on sets from which all required operations may be constructed through composition and application. Category theory provides some guidance on this matter (see chapter 7). We will often display finite sets using the standard set-parentheses {...}.

In the definition of the abstract type of finite sets, the clause for the extraction of an element from the empty set raises an exception. Exceptions are the topic of the next section.

2.8 Exceptions

Often when programming we meet expressions which are not intended to have any value, for instance when a function is applied to an argument outside its domain of definition. To ensure that programs are 'robust' in being able to cope with erroneous input, we need a so-called exception mechanism. There is a type-safe exception mechanism in ML which we illustrate with an example:

```
exception empty_list: unit

fun head(nil)  = raise empty_list
  | head(a::s) = a
```

The name of the exception empty_list and its type unit is declared. The type of an exception is that of the value returned (in this case no value is returned). The function evaluates as follows. When the list is not empty, this function returns the first item of the list. When it is empty, evaluation ceases and an exception with name empty_list is raised. When an exception is raised, evaluation may be passed to another expression called the *handler* as in the following, rather contrived, function for concatenating lists end to end.

```
fun append(s,t) =
      head(s)::append(tail(s),t) handle empty_list => t
```

Here, the first part of the definition clause is evaluated. If no exception is raised, its value is returned. If an exception called empty_list is raised, the value of the clause in the handler is returned. To handle all exceptions raised in a clause we use a 'wildcard' instead of an exception name: ... handle ? => ....

Values may be passed through exceptions as in the following example:

```
exception div_by_zero:int
fun divide(n,d) = if d=0
        then raise div_by_zero with n else div(n,d)
```

An expression calling this function can make use of the value of the numerator when the function fails:

```
f(a,b) handle div_by_zero with x => x*x
```

There are certain scope rules associated with exceptions which are explained in [Harper, Mitchell 86].

2.9 Other facilities

ML provides a facility for modular programming, based on a proposal of MacQueen [1985]. It incorporates parameterized modules and submodule sharing. We do not describe modules here as they do not appear in the forthcoming programming (for reasons discussed in Chapter 10).

Other features of ML are a type-safe reference and assignment facility and a collection of input/output primitives based on streams.

2.10 Exercises

The exercises in this chapter are designed to be done interactively on an ML system. They may, however, be done as pen-and-paper exercises. They follow roughly the order of presentation in the chapter, with some more substantial exercises at the end. Answers to the exercises are given in Appendix D.

Exercise 1. Values and environments What values or environments are given by the following ML expressions?

1. `val x = 3; val y = 4 and z = x+1`
2. `let val x =1 and y =2 in x+y end`
3. `val p = 3 and q = p+1`
4. `let val (x,y) = (2,3) in 2*x + y end`
5. `let val x = 1 in let val y = x+2 in let val x = 5 in x+y end end end`
6. `val (x,y as (_,p)) = ((2,3),(4,(5,6)))`

Exercise 2. Defining functions Define the following functions on
integers:

1. The function `sign` which tests whether an integer is positive.

2. The function `absvalue` which returns the absolute value of an
   integer.

3. The function finding the maximum of two integers.

4. The Fibonacci sequence is $1, 1, 2, 3, 5, 8, 13, \ldots$ in which each
   number is the sum of its two immediate predecessors. Write
   a recursive definition of the n-th entry in the sequence.

Exercise 3. Natural numbers Define the type of natural numbers as
follows:

```
datatype Num = zero | succ of Num
```

Define a function `numprint :  Num -> int` which displays natural
numbers as integers.

In the text of the chapter we show how to define addition of
natural numbers. Use this addition operation and the same cases
as in its definition to define the multiplication of natural numbers.

Exercise 4. Higher order and polymorphic functions What are
the most general types of the following functions?

1. The function `apply` which takes a function and a value and
   returns the results of applying the function to the value:

   ```
   fun apply(f)(x) = f(x)
   ```

2. The function which composes two functions:

   ```
   fun compose(g,f) = fn x => g(f(x))
   ```

Exercise 5. List processing Define the following functions on lists.

1. The function which finds the maximum integer in a list of
   integers.

2. The function which sums a list of integers.

3. The function which takes a list of coefficients $a_0, a_1, \ldots, a_n$ and
   a value x and evaluates the polynomial $a_0 + a_1 \times x + \ldots a_n \times x^n$.

4. Use the `append` function, concatenating lists end to end, to define the function which reverses a list.

5. The function `maplist` which applies a function to all items in a list returning the list of results. What is its most general type?

6. The function calculating the sum of a list of integers can be generalized. Suppose there is a binary function `f:  A*B ->` B and an initial value `v:B`, then we may run through an A `list` accumulating a result by successively applying the binary function to the current element of the list and the value accumulated so far, starting with the initial value. Define this function – the definition is shorter than its explanation!

Exercise 6. Binary trees For this exercise, a binary tree is a structure like:

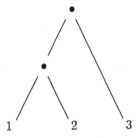

It consists of binary branching nodes and values stored at the tips. The top node is called the root.

As a type within ML, we define binary trees as follows:

```
datatype 'a BinTree =
    tip of 'a | node of ('a BinTree)*('a BinTree)
```

The tree above is then represented as the expression:

```
node(node(tip(1),tip(2)),tip(3))
```

Define the following functions on binary trees:

1. The breadth of a tree, defined as the number of tips.

2. The depth of the tree, defined as the maximum length of a path from the root to a tip.

3. The function which collects, in order, the list of values at the tips.

Exercise 7. Data abstraction Rational numbers may be represented as integer fractions or as an integer part together with a fractional part which consists of a finite sequence of digits followed by another finite sequence which is infinitely repeated. Define the arithmetic of rational numbers as an abstract type using either representation.

Exercise* 8. More list processing We have already seen many functions on lists. However, this is but a small sample of a rich vein of functions which illustrate the utility of recursion as a concise and executable definition mechanism. Here are a few more suggestions for list processing functions. Clearly there are many similar functions which you may wish to encode.

1. The function which deletes all occurrences of a value from a list is defined as follows:

```
fun delete(x,nil) = nil
  | delete(x,a::s) =
        if x=a then delete(x,s)
               else a::delete(x,s)
```

Define the function which deletes the n-th occurrence of a value.

2. Define the function `sublist` which tests whether a list is a sublist of another, in the sense that the second list is the first filled out at any positions with other entries. There are variants of this notion of sublist which you may like to encode.

3. Define the function which counts how may times a list is a sublist of another, including overlapping. Here are some sample results.

```
number_of_sublists([1,2],[1,2,2])
> 2 : int
number_of_sublists([1,2],[1,1,2,2])
> 4 : int
number_of_sublists([1,1],[1,1,1])
> 3 : int
number_of_sublists([1,1],[1])
> 0 : int
```

This is rather tricky so we give the first clauses in the definition:

```
fun number_of_sublists(nil,t) = 1
  | number_of_sublists(a::s,nil) = 0
  | number_of_sublists(a::s,b::t) = ...
```

Exercise* 9. Operations on finite sets In this exercise we use the abstract type of finite sets defined in Section 2.7 and define several operations on sets. These operations will arise in the following chapters, when we consider internal structure within the category of finite sets.

Let us begin with a definition of a function which takes the image of a finite set through a function:

```
fun image(f)(s) = if is_empty(s) then emptyset else
          let val (x,s') = singleton_split(s) in
              union(singleton(f(x)),image(f)(s')) end
```

Now define the following operations:

1. The disjoint union of finite sets.
2. The cartesian product of two finite sets.
3. The powerset of a finite set – the set of all subsets of the set.
4. The set of all total functions between two finite sets. For this, represent a function between two finite sets as its graph (list of argument-result pairs).

Notice how the definitions of the last three functions depend upon one another. This dependence can be expressed abstractly within category theory.

To print sample results, you will need a function converting sets to strings of characters for display. Moreover, to define powersets you need equality on finite sets.

Exercise* 10. Sorting This is an exercise in programming an algorithm for sorting a list of items, which support a total order, into a non-descending sequence. The algorithm is called 'tree sort' and works by inserting items successively into an 'ordered' tree and then flattening the resultant tree. Those unfamiliar with the algorithm should consult a reference such as [Knuth 73].

The algorithm uses binary trees of the following form (where we consider, for simplicity, only sorting lists of integers):

```
datatype BTree = empty | tip of int |
                 node of BTree*int*BTree
```

We build ordered trees. A tree `node(s,n,t)` is ordered if the nodes in s are all less than n and those in t are greater than or equal to n and s and t are ordered.

Write a function `insert` for inserting a value in an ordered tree by creating a new node and still maintaining the ordered property. Also write a function `flatten` to collect the list of values at the nodes following an in-order traversal.

Using the function `accumulate` defined in Exercise 5.6 above, the sorting algorithm can be expressed as:

```
fun sort(s) = flatten(accumulate(insert)(empty)(s))
```

Exercise* 11. Universal algebra and recursion Burstall and Landin [1969] show how ideas from universal algebra can contribute to the design of computer programs. This is a short exercise based on these ideas and should only be attempted if you know some universal algebra.

Notice that the definition of list processing functions like `length`, `member` (Section 2.6), `sum` and `maplist` (Exercise 5) all have the same general form. This may be explained by the fact that lists form a free monoid. The monoid structure is that of concatenating lists with the empty list as the identity. There is a function $h : A \rightarrow list(A)$ which returns the singleton list on an element. The freeness is expressed by the unique existence of a homomorphism as follows:

For any monoid $(B, *, e)$ and function $f : A \rightarrow B$, there is a unique homomorphism $f^\#$ from the monoid of lists to $(B, *, e)$ such that the following commutes:

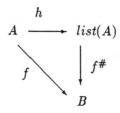

Given the monoid $(B, *, e)$, the map $f \mapsto f^\#$ can be constructed and the construction can be expressed as a program in ML. Representing the target monoid as a pair of a binary function and a constant, write this program. It is an example of an encapsulation of recursion, in that functions (like those mentioned above) normally defined through explicit recursion, can be obtained by passing suitable parameters to this program.

Consult the reference above for more details of this and for the extension from lists to arbitrary free algebras.

Chapter 3

Categories and Functors

Beginning in this chapter and running through to Chapter 7, we develop some of the basic ideas in category theory. Examples are chosen for relevance to computing. Theorems are established through constructive proofs. The presentation of the mathematics is accompanied by corresponding ML programs.

The starting point in the development of category theory is the definition of a category together with illustrative examples. We present this material much as it is to be found in standard texts, concentrating however on examples of relevance to programming. Alongside this, we begin the programming of category theory by representing categories so as to compute with them.

3.1 Categories

Category theory is founded upon the abstraction of the arrow,

$$f : a \to b$$

Here a and b are called objects and f is an arrow whose source is object a and target is object b. Such directional structures occur widely in set theory, algebra, topology and logic. For example, a and b may be sets and f a total function from a to b or, indeed, f may be a partial function from set a to set b; or a and b may be algebras of the same type and f a homomorphism between them; or a and b may be topological spaces and f a continuous map; or, again, a and b may be propositions and f a proof of $a \vdash b$.

It is by describing structure in terms of the existence and properties of arrows that category theory achieves its wide applicability. The usual mode of description in mathematics is by reference to the internal structure of objects. The applicability of the description is then limited to objects supporting such structure. Categorical descriptions make no assumption about the internal structure of objects; they are purely in terms of the 'transport' of whatever structure is preserved by the arrows. In this sense, they are data independent descriptions – the same description may apply to sets, graphs, algebras and whatever else can be considered to be objects in a category.

Particularly amenable to description in terms of arrows are constructions which are in some sense 'canonical'. These are common throughout mathematics. We mention a few here to illustrate the sorts of constructions we have in mind. Canonical constructions in graph theory are the transitive closure of a graph and the strong components of a graph. In algebra, free and generated algebras are common. A canonical construction is the abelianization of a group. In topology, there are constructions like the compactification of spaces. An arrow-theoretic description of such constructions captures all the ingredients, including the sense in which the construction is considered to be canonical.

The generality of descriptions in term of arrows is offset by a remoteness from application so that considerable work is involved in unravelling categorical descriptions in a particular setting and, conversely, in attempting to give a categorical description of a particular concept. On the other hand, these descriptions are usually elementary (i.e. first-order) though they tend to be fairly complex in terms of the alternation of quantifiers.

To support definitions in terms of arrows, Eilenberg and Mac Lane [1945] introduced structures called categories. A category is a class whose elements are 'objects' together with a class of 'arrows' (sometimes called 'morphisms') between objects. Arrows are to be composable: if $f : a \to b$ and $g : b \to c$, there is a composite arrow $gf : a \to c$ (sometimes denoted $g.f$). Notice the order in which we write the composition. Some authors write fg rather than gf to denote the composite of $f : a \to b$ followed by $g : b \to c$. The order fg corresponds to the diagram,

$$a \xrightarrow{f} b \xrightarrow{g} c$$

but then, to be consistent, application of a function f to an argument x becomes post application xf (like field selectors in some programming languages). The order gf corresponds to the usual prefix notation for the application of functions $(gf)(x) = g(f(x))$. The latter is gaining ascendancy in categorical texts and so is adopted here.

Composition is to have two properties:

1. Associativity. For all $f : a \to b$, $g : b \to c$ and $h : c \to d$,

$$(hg)f = h(gf)$$

2. Identity. For all objects a in the category there is an 'identity' arrow i_a such that for all $f : a \to b$,

$$fi_a = f = i_b f$$

Categories are thus graphs (directed multigraphs) with a composition and identity structure. Based upon this, we may give a formal definition of a category.

Definition 1 *A graph is a pair N, E of classes (of* nodes *and* edges*) together with a pair of mappings $s, t : E \to N$ called* source *and* target *respectively. We write $f : a \to b$ when f is in E and $s(f) = a$ and $t(f) = b$.*

Definition 2 *A* category *is a graph (O, A, s, t) whose nodes O we call* objects *and whose edges A we call* arrows. *Associated with each object a in O, there is an arrow $i_a : a \to a$, the* identity *arrow on a, and to each pair of arrows $f : a \to b$ and $g : b \to c$, there is an associated arrow $gf : a \to b$, the* composition *of f with g. The following equations must hold for all objects a, b and c and arrows $f : a \to b$, $g : b \to c$ and $h : c \to d$:*

$$(hg)f = h(gf)$$

$$fi_a = f = i_b f$$

Such then are categories. The structure is fairly weak, consisting of only typed composition with identities.

It is customary to say something about the 'size' of the collections O and A which we have called classes. Many categories have collections O and A which are too 'big' to be sets, for instance, the category whose objects are all sets (in some universe) or whose objects are all groups (again in some universe). Considering these collections to be sets yields inconsistency (Russell's paradox). We thus need some justification for dealing with these large categories in terms of a foundational framework for category theory. Originally, various extensions of set theory were proposed to deal with this matter. Universes of sets were considered by Mac Lane [1971], Grothendieck [1963] and Feferman [1969] and the Gödel-Bernays theory of classes has also been employed as a foundation. Set-theoretic foundations are discussed in Blass [1984]. When working in category theory, such set-based foundations seem both superfluous (except to avoid contradictions) and inappropriate. There have been proposals that category theory itself should provide a foundation not only for category theory but also for the whole of mathematics. Lawvere [1966] considers the category of categories as a foundation, whilst Bénabou [1985] analyses category theory in use and proposes fibered categories as a foundation (to handle 'set'-indexed diagrams).

Programming category theory, as we are about to do, involves embedding it in a suitable formal language. In doing so we confront this foundational question. From a programming point of view, we are interested only in objects and arrows that we may construct. This leads, as we shall see (Section 3.3), to a framework based on formal type systems. Various systems have been proposed and some have been incorporated in programming languages. We discuss this topic in Chapter 10.

3.1.1   Diagram chasing

Categorical properties are often expressed in terms of 'commuting' diagrams and proofs take the form of 'diagram chasing'. Informally, a diagram is a picture of some objects and arrows in a category. Formally, a diagram is a graph whose nodes are labelled with objects of the category and whose edges are labelled with arrows in the category in such a way that source and target nodes of an edge are labelled with source and target objects of the labelling arrow. An example of a finite diagram Δ is:

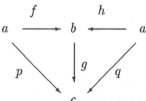

A *path* in a diagram is a non-empty sequence of edges and their labelling arrows such that the target node of each edge is the source node of the next edge in the sequence. For example, the left triangle in the diagram Δ contains the path, $a\xrightarrow{f}b\xrightarrow{g}c$. Each path determines an arrow by composing the arrows along it. A diagram is said to *commute* if, for every pair of nodes m, n, every path from m to n determines through composition the same arrow. Thus to say the diagram Δ commutes amounts to the following two equations:

$$p = gf \quad \text{and} \quad q = gh$$

Proofs using commutation of diagrams are equational proofs. The equations are *typed* by their source and target nodes. This simple form for categorical proofs means that these proofs are especially amenable to be conducted on computers. Not only is the technology for equation deduction fairly well understood but also the typed nature of categorical proofs constrains the search arising in deduction. Huet [1986] has considered this matter and Watjen and Struckmann [1982] have automated part of categorical reasoning.

3.1.2 Subcategories, isomorphisms, monics and epis

A category **C** whose objects and arrows are subclasses of those of a category **A**, and whose source, target, identities and compositions are those of **A** is said to be a *subcategory* of **A**. A subcategory **C** of **A** is said to be *full* when, for all pairs of objects a and b in **C**, if $f : a \to b$ is an arrow in **A** then it is in **C**. A full subcategory of a category **A** is determined by the objects alone.

We now consider some important properties of arrows. An arrow $f : a \to b$ in a category is said to be an *isomorphism* if there is an arrow $g : b \to a$ which is a left and right inverse to f i.e. $gf = i_a$ and $fg = i_b$. In this case we say that a and b are *isomorphic* objects, written $a \cong b$. In the category whose objects are sets and whose arrows are functions, isomorphisms are bijections (1-1 and onto functions). Isomorphisms are important in category theory since arrow-theoretic descriptions usually determine an object only to within an isomorphism. Thus isomorphisms are the degree of 'sameness' that we wish to consider in categories.

An arrow $m : b \to c$ is a *monic* iff for every pair of arrows $f, g : a \to b$, if $mf = mg$ then $f = g$. By turning the arrows around, we define the concept of an epi: an arrow $e : a \to b$ is a *epi* iff for every pair of arrows $f, g : b \to c$, if $fe = ge$ then $f = g$. In the category of sets with functions as arrows, monics are 1-1 functions and epis are onto functions. In fact, monics are but one of several different categorical notions which characterize 1-1 functions, likewise epis and onto functions.

3.2   Examples

Examples of categories abound throughout mathematics. We consider some that are of importance in programming.

3.2.1   Sets and finite sets

The primary example of a category is that of sets, **Set**, whose objects are sets (in a universe) and whose arrows are total functions. More precisely, since arrows determine their source and target objects, arrows in **Set** are *typed* total functions, which we may consider to be triples (a, f, b), a and b being sets and f a function defined on all elements of a and whose results lie in b. Identities are simply identity functions: $i_a = (a, \lambda x.x, a)$ and composition is the composition of functions, for $f : a \to b$ and $g : b \to c$,

$$gf = (a, \lambda x.g(f(x)), c)$$

A subcategory of **Set** is that of finite sets, **FinSet**, whose arrows are again typed total functions.

There are other categories whose objects are sets. For instance, we may consider arrows to be not total functions but partial functions to get a category $\mathbf{Set}_{Pf}$. We may go further and consider objects again to be sets but arrows to be relations between sets (labelled with their source and target sets). A relation $r : a \to b$ is a subset of the cartesian product $a \times b$. The composition sr of $r : a \to b$ with $s : b \to c$ is defined by

$$sr = \{(x, z) : \exists y \in b \,.\, (x, y) \in r \land (y, z) \in s\}$$

Let us call this category $\mathbf{Set}_{Rel}$.

3.2.2   Graphs

We consider directed multi-graphs, that is, pairs of sets N (of *nodes*) and E (of *edges*) together with pairs of functions $s, t : E \to N$ (*source* and *target* respectively). Notice that the collections of nodes and of edges are both sets rather than proper classes (so-called *small* graphs).

Graphs are objects in a category **Graph**. Arrows in this category are structure-preserving maps between graphs. This means that an arrow from graph (N, E, s, t) to graph (N', E', s', t') is a pair of functions $(f : N \to N', g : E \to E')$ such that, for all $e \in E$, $f(s(e)) = s'(g(e))$ and $f(t(e)) = t'(g(e))$. We may picture these two requirements as the commutation of the following squares:

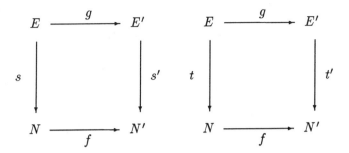

3.2.3 Finite categories

Categories in which the classes of objects and of arrows are both finite sets may be represented pictorially. For example,

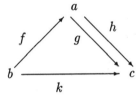

Here a, b and c denote objects, f, g and h are arrows with sources and targets as depicted. We define composition explicitly for each composable pair of arrows: $gf = k$ and $hf = k$ and the existence of an identity arrow for each object is assumed.

3.2.4 Relations and partial orders

We have already defined the category $\mathbf{Set}_{Rel}$ whose objects are sets and whose arrows are relations. However, we can also define a category whose *objects* are relations. To do so we must say what we mean by an arrow between relations.

First some notation: if $R \subseteq a \times b$ is a relation we will write xRy for $(x, y) \in R$. Now consider relations from a set to itself. These are pairs $(a, R \subseteq a \times a)$. An arrow from (a, R) to (b, S) is a function $f : a \to b$ such that $xRy \Rightarrow f(x)Sf(y)$. Composition and identities in this category, **Rel**, are those of functions.

A subcategory of **Rel** is the category of partial orders. A relation $(a, \leq)$ is a partial order if it satisfies the following:

1. (Reflexivity) $x \leq x$.

2. (Anti-symmetry) If $x \leq y$ and $y \leq x$, then $x = y$.

3. (Transitivity) If $x \leq y$ and $y \leq z$, then $x \leq z$.

Arrows between partial orders are arrows of relations, i.e. order preserving functions. This forms a category **Pos**.

3.2.5 Partial orders as categories

As well as the category **Pos** of partial orders, each partial order $(a, \leq)$ may itself be considered to be a category as follows. The objects of the category are the elements of a. There is precisely one arrow from $x \in a$ to $y \in a$ iff $x \leq y$. Transitivity ensures that composition is defined and reflexivity ensures that identities exist. In fact, we do not need the anti-symmetry axiom, so this construction works for *pre-orders* (reflexive, transitive relations).

3.2.6 Deductive systems

The idea here is that propositions are objects in a category and that an arrow $f : a \to b$ corresponds to (an equivalence class of) a proof of $a \vdash b$.

Let us illustrate with a simple example: The calculus of conjunction (logical 'and' denoted $\wedge$). The objects are propositional expressions built from variables (drawn from a set X say) and a binary infix operator denoted $\wedge$.

The arrows are equivalence classes of proofs generated by inference rules. We write the inference rules using (proofs of) sequents rather than propositions, so that inference rules provide means of constructing arrows.

Corresponding to the deduction $a \wedge b \vdash a$, we have the rule

$$\pi_{a,b} : a \wedge b \to a$$

and likewise the rule

$$\pi'_{a,b} : a \wedge b \to b$$

These are *elimination* rules for conjunction. The *introduction* rule, constructing arrows from arrows, is

$$\frac{f : c \to a, \quad g : c \to b}{\langle f, g \rangle : c \to a \wedge b}$$

To make this structure into a category, we assume the existence of identity entailments $a \vdash a$:

$$i_a : a \to a$$

and an associative composition of proofs,

$$\frac{f : a \to b, \quad g : b \to c}{gf : a \to c}$$

for which the identity arrows are indeed identities. Finally, we impose the following equations on proofs:

$$\pi_{a,b}\langle f, g \rangle = f$$

$$\pi'_{a,b}\langle f, g \rangle = g$$

$$\langle \pi_{a,b}h, \pi'_{a,b}h \rangle = h$$

As an example proof, we demonstrate the commutativity of $\wedge$, i.e. $a \wedge b \vdash b \wedge a$. The following arrow is a proof of this entailment:

$$\langle \pi'_{a,b}, \pi_{a,b} \rangle : a \wedge b \to b \wedge a$$

This is but an indication of how deductive systems give rise to categories. Further structure in the logic is reflected in corresponding categorical structure. There is now a well developed field of categorical logic, including topos theory (Chapter 7) and structures combining the idea of proofs as arrows with that, introduced in the next section, of term substitutions as arrows. For more details, the reader should consult references such as [Goldblatt 79], [Lambek, Scott 86], [Lawvere 70], [Seely 83].

3.2.7   Universal algebra: terms, algebras and equations

We describe some basic universal algebra and several categories which occur. For more details of universal algebra the reader should consult standard texts, for example [Cohn 65,81]. Some computational aspects of universal algebra are dealt with in [Huet 80] and [Huet, Oppen 80]. Indeed the concepts developed here are fundamental to the semantics of computation [Goguen, Meseguer 88].

The starting point is the notion of a 'term' which is a symbolic expression built from operators (function symbols) and variables. Thus if $X = \{x, y, z\}$ is a set of variables (simply symbols) and f is a unary operator and g a binary operator, then the expressions $f(f(z))$ and $g(f(x), g(x,y))$ are terms in X.

An *operator domain* is a set of operator symbols indexed by their arities (natural numbers). If Ω is an operator domain, denote by Ω_n the set of operators in Ω whose arity is the natural number n.

The *terms* in a set X over an operator domain Ω, the set of which we denote by $T_\Omega(X)$, are syntactic objects defined recursively by:

$$x \in X \Rightarrow \langle x \rangle \in T_\Omega(X)$$

$$\rho \in \Omega_n, \ t_1, t_2, \ldots, t_n \in T_\Omega(X) \Rightarrow \rho(t_1, t_2, \ldots, t_n) \in T_\Omega(X)$$

Strictly speaking, a constant c, i.e an operator with arity 0, when considered as a term should be written $c()$ but we will often elide this distinction. Also variables should be enclosed as $\langle x \rangle$ but the parentheses are dropped where no ambiguity arises.

A *(term) substitution* from set X to set Y, $f : X \to Y$ is a function, $f : X \to T_\Omega(Y)$, mapping variables to terms. Thus if $X = \{x, y, z\}$ and $Y = \{u, v, w\}$ and f is a unary operator and g a binary operator, then an example term substitution from X to Y is:

$$\{x \mapsto g(u, v), y \mapsto g(f(u), g(u, v)), z \mapsto f(f(w))\}$$

We now define a category whose objects are sets and whose arrows are term substitutions. To do so we define composition of term substitutions and identity substitutions.

Notice that substitutions can be applied to terms. If $f : X \to Y$ is a substitution, define the application of f to a term in X by:

$$f(\langle x \rangle) = f(x) \text{ for all } x \in X$$

$$f(\rho(t_1, t_2, \ldots, t_n)) = \rho(f(t_1), f(t_2), \ldots, f(t_n))$$

$$\text{for } \rho \in \Omega_n, \ t_1, t_2, \ldots, t_n \in T_\Omega(X)$$

We define composition using application $(gf)(x) = g(f(x))$. Also, for each set X the identity substitution is defined by $i_X(x) = \langle x \rangle$ (unless X is empty in which case the identity is the empty function).

Define $\mathbf{T}_\Omega$ to be the category whose objects are sets and whose arrows are substitutions. This is indeed a category under the composition and identities above. The full subcategory of finite sets is denoted $\mathbf{T}_\Omega{}^{Fin}$.

This category provides a basis for a categorical treatment of equational deduction. In Chapter 8, we show how a particular case of equation-solving, the so-called unification problem, can be interpreted in this category and we derive unification algorithms from general categorical constructions. For further details of this categorical treatment of equational deduction, see [Rydeheard, Stell 87].

We now consider algebras. An algebra is a set together with a family of functions on the set. By introducing arrows (homomorphisms) between algebras of the same kind, we may define categories of algebras.

An Ω-algebra (A, α) consists of a set A called the *carrier*, together with, for each $n \in \mathcal{N}$ and operator $\rho \in \Omega_n$, a function $\alpha_\rho : A^n \to A$. A homomorphism h from (A, α) to (A', α') is a function $h : A \to A'$ such that for each $n \in \mathcal{N}$ and $\rho \in \Omega_n$, and $x_1, x_2, \ldots, x_n \in A$,

$$h(\alpha_\rho(x_1, x_2, \ldots, x_n)) = \alpha'_\rho(h(x_1), h(x_2), \ldots, h(x_n))$$

Composition and identities are the composition and identities of functions. Thus, for each operator domain Ω, we form the category of Ω-algebras, $\mathbf{Alg}_\Omega$.

As an example, we may take Ω to contain just a nullary operator (one taking no arguments, i.e. a constant), c, and a unary (1-argument) operator, f. An Ω-algebra consists of a set A, a distinguished element of A as the constant and a unary function on A. The natural numbers $\mathcal{N} = \{0, 1, \ldots\}$ form an Ω-algebra with c as 0 and f as the successor function, $\lambda x . x + 1$. Another Ω-algebra is the set of truth-values $\mathcal{T} = \{\top, \bot\}$ with c as $\top$ and f defined to be negation $\lambda x . \neg x$. There is a homomorphism $h : \mathcal{N} \to \mathcal{T}$ given by $h(0) = \top$ and $h(x + 1) = \neg(h(x))$. This is the function which tests whether a number is even. Notice that these equations not only define the function but also ensure that it is a homomorphism.

Groups and rings are examples of algebras whose operators satisfy a set of equations. This set of equations is often called the 'theory' of groups or rings respectively. The class of all groups, and that of all rings, is the class of algebras which satisfy the equations. This class is called a 'variety'. By introducing homomorphisms as arrows, varieties become categories – full subcategories of $\mathbf{Alg}_\Omega$ for suitable Ω.

Let us state this formally. First notice that the set $T_\Omega(X)$ may be considered to be an Ω-algebra $(T_\Omega(X), \tau)$ by defining the operators to construct terms, i.e. for each $n \in \mathcal{N}$ and $\rho \in \Omega_n$:

$$\tau_\rho(t_1, \ldots, t_n) = \rho(t_1, \ldots, t_n)$$

Moreover, there is a function $h : X \to T_\Omega(X)$ defined by $h(x) = \langle x \rangle$.

Given an algebra (A, α), a function $f : X \to A$ determines a homomorphism $f^\# : (T_\Omega(X), \tau) \to (A, \alpha)$ defined by

$$f^\#(\langle x \rangle) = f(x)$$

$$f^\#(\rho(t_1, t_2, \ldots, t_n)) = \alpha_\rho(f^\#(t_1), f^\#(t_2), \ldots, f^\#(t_n))$$

In fact this homomorphism is the unique homomorphism such that $f^\# h = f$. This is described by saying that $(T_\Omega(X), \tau)$ is a *free* Ω-algebra on the set X. Freeness and this construction of free algebras is discussed again in Chapter 6.

An Ω-equation is a pair of terms $\langle t_1, t_2 \rangle$. An Ω-algebra (A, α) *satisfies* an equation $\langle t_1, t_2 \rangle$ in variables X if, for all functions $f : X \to A$, $f^\#(t_1) = f^\#(t_2)$.

Let $\mathcal{E}$ be a set of Ω-equations. The pair $\mathbf{T} = (\Omega, \mathcal{E})$ is called a presentation of an equational theory (or just a theory). An Ω-algebra is a $\mathbf{T}$-algebra if it satisfies all the equations in $\mathcal{E}$. We may thus define the category of all $\mathbf{T}$-algebras with homomorphisms as arrows, $\mathbf{Alg_T}$.

The category of groups, $\mathbf{Group}$, is an example of an equational variety. Let Ω contain a nullary operator e, a unary operator i and a binary (infix) operator $\circ$. The equations $\mathcal{E}$ are the associativity of $\circ$, the left and right identity laws for e and the left and right inverse laws for i. If $\mathbf{T} = (\Omega, \mathcal{E})$, then $\mathbf{Group} = \mathbf{Alg_T}$.

It is possible to consider constraints other than equations, e.g. Horn clauses or formulae in first order logic, to define full subcategories of the categories $\mathbf{Alg_\Omega}$.

So far, algebras contain just one set and operations on it. More generally, we may consider 'many sorted' algebras which contain more than one set. These arise as data types in programming and are discussed in Chapter 9.

For those with knowledge of universal algebra, we mention that $\mathbf{T_\Omega}$ is equivalent to the category of free Ω-algebras with homomorphisms as arrows. In fact, the construction of $\mathbf{T_\Omega}$ is a special case of a general construction due to Kleisli [1965]. For details of this, consult [Manes 76], [Mac Lane 71] or [Barr, Wells 85] or the related 'algebraic theories' of Lawvere [1963].

3.2.8 Sets with structure and structure-preserving arrows

This is but a small selection of categories which arise. Many categories have objects which are sets with structure upon them and arrows which are functions preserving structure, with identities as identity functions and composition as function composition. We have already met the categories **Graph**, **Rel** and **Alg**$_\Omega$ which are of this form. The preservation of structure may be the mapping of structure in the source to a corresponding structure in the target. Other forms of preservation are present in topology where there are categories **Top**, of topological spaces and continuous maps, **Top**$_{Open}$, of topological spaces and open maps, **Haus**, the category of Hausdorff spaces and continuous maps, amongst many others.

3.3   Categories computationally

We now look at categories from a programming viewpoint. Our aim is to capture the computational content of the definition of a category so that constructions within categories may be expressed as computer programs.

What then is the computational content of a category? The objects in a category are delimited by their common structure. This common structure is captured by functions which build objects from other structures and, possibly, from other objects in the category. A class of values described in terms of functions for constructing values is the notion of a 'type' in programming. Thus the class of all objects and, likewise, that of all arrows, is captured by a type within a suitable programming language. Types in programming languages are not always sufficiently expressive to define these classes. This is particularly so for languages without explicit types like Lisp. Nevertheless, the common structure of objects and arrows is captured by the form of the representation which can be expressed as functions for constructing representations.

The category structure on the objects and arrows is represented again using functions. Four functions are required to specify how the objects and arrows are considered to form a category. These functions are the source and target functions, returning an object for each arrow, the identity function, returning an arrow for each object, and the composition function which, for two composable arrows, returns their composite arrow. These functions together with the type of the objects and of the arrows define a category and contain all its basic computational content.

Rarely do we consider categories without some additional internal structure. For example, categories may be equipped with products or coproducts of pairs of objects. We shall define these constructs in the next chapter and show that they can be represented as functions. Thus a category with additional structure consists of the four functions mentioned above together with extra functions recording the additional structure. Notice how we turn *properties* of a category (e.g. that all pairs of objects have a product) into functions whose existence is the property in question.

It is important to capture in the programming the generality of categorical constructions. The constructions apply to a range of categories. To compute a result, the construction is specialized to a particular category. This is achieved by binding identifiers in the program to the functions defining the category. The construction can then call upon these functions in its computation. This describes the passing of categories as parameters to programs. We see that a higher-order capability, passing functions as parameters, is required to program categorical constructions. This may be either implicit, as in some algebraic systems, or explicit as in ML.

Many different types may be the objects and arrows of categories. We have defined a category **Set** whose objects are sets, a category **Graph** whose objects are graphs and a category $\mathbf{Alg}_\Omega$ whose objects are algebras. A categorical construction applies to a range of categories quite possibly with different types of objects and arrows. The construction is uniform in that the same program is used for the computation in different categories, the difference arises only in the types and the evaluation of the identity, composition and other functions defining the category. This phenomenon of a program applying to values of various types, and hence the function computed having a range of types, is the concept of (parametric) polymorphism discussed in the previous chapter.

The programming language ML, which we use here, allows both higher-order functions (functions taking functions as arguments or returning them as results) and a form of type polymorphism. Our immediate aim is to express categories as a type in ML and then write code for particular categories to yield values of this type.

Recall that a category is described in terms of two types capturing the common structure of the objects and the arrows, along with four functions defining the category structure on the objects and arrows. These functions are the source, target, identity and composition functions. If `'o` is the type of the objects, and `'a` the type of the arrows, then the source and target functions both have type `'a -> 'o`, the identity function has type `'o -> 'a` and composition function has type `'a*'a -> 'a`. We thus declare the type of categories to be that of 4-tuples of functions with these types:

```
datatype ('o,'a)Cat =
    cat of ('a->'o)*('a->'o)*('o->'a)*('a*'a->'a)
```

This type declaration defines a type constructor Cat which, when supplied with types for the objects and arrows gives the type of all categories with objects and arrows of the specified type. The right-hand side of the declaration states that values of the type are 4-tuples of functions with the given types. This 4-tuple is labelled with the data constructor cat – an uninterpreted function turning 4-tuples of functions into categories. Whether categories are labelled or unlabelled tuples is, from a computational viewpoint, merely a matter of taste, for categories are all of one form. This should be contrasted with other types, e.g. lists where the labels nil and :: are essential for distinguishing the two forms of lists. We choose to label the values so as to record which tuples are to denote categories and to document the fact that these tuples of functions satisfy some axioms defining a category.

3.4 Categories as values

To define a category we need the following: types to represent the objects and arrows, and definitions of functions for the source and target of arrows, the identity arrows and the composition of arrows. We look at some examples of categories as values in ML.

3.4.1 The category of finite sets

Let us consider the category **FinSet** of finite sets and typed total functions. To begin, we need a type of finite sets. In some languages, like Pascal, finite sets are implemented as a basic type. In languages where this is not the case we represent finite sets using other types such as arrays or lists. In the previous chapter we defined sets as an abstract type represented by linear lists. Using this representation, we denote by 'a Set the type of finite sets whose elements have type 'a.

Arrows in **FinSet** are typed functions – triples consisting of two sets and a function between them. As a data type this is:

```
datatype 'a Set_Arrow =
    set_arrow of ('a Set)*('a->'a)*('a Set)
```

We now define four functions describing how finite sets and set arrows form a category.

```
fun set_s(set_arrow(a,_,_)) = a
fun set_t(set_arrow(_,_,b)) = b
fun set_ident(a) = set_arrow(a,fn x => x,a)
fun set_comp(set_arrow(c,g,d),set_arrow(a,f,b)) =
    if seteq(b,c) then set_arrow(a,fn x => g(f(x)),d)
                  else raise non_composable_pair
```

The notation **fn x => e** is that of the λ-expression $\lambda x.e$.

It is at this point that we consider the partial nature of composition. Composition is defined not on all pairs of arrows but only on composable pairs, those for which the target object of the first is the source object of the second. To test whether this is indeed the case requires computing equality on objects in the category. Here we test whether two finite sets are equal using the function **seteq** which, for two finite sets a and b, tests whether the following holds:

$$(\forall x \in a . \exists y \in b . x = y) \wedge (\forall y \in b . \exists x \in a . y = x)$$

Thus equality on finite sets requires an equality function on the elements. In ML, some types, including primitive types, support an equality by virtue of the structure of their values. Types which include functions as part of their values do not have an implicit equality. If an equality is required in this case then it must be given explicitly.

Notice that an attempt to compose two non-composable arrows is signalled by raising an exception. In this case all that happens is that the token message **non_composable_pair** is returned. More informative error handling could be included if required.

These functions on sets and set arrows are bundled together to form the category of finite sets, which is a value of type,

```
('a Set,'a Set_Arrow)Cat
```

and is defined by:

```
val FinSet = cat(set_s,set_t,set_ident,set_comp)
```

3.4.2   Terms and term substitutions: the category $\mathbf{T}_\Omega{}^{Fin}$

We look at a computational representation of one of the categories arising in universal algebra discussed in Section 3.2.7. We consider the category $\mathbf{T}_\Omega{}^{Fin}$ whose objects are finite sets and whose arrows are term substitutions.

We define operator domains, terms and term substitutions in ML. An operator is an operator symbol together with an arity. We take arities to be sets of names of argument places rather than just natural numbers, so that we identify the argument places in a term $\rho(t_1, \ldots, t_n)$ with the set of numbers $\{1, \ldots, n\}$. More generally the argument places can be names, for example, the division operator on numbers may be denoted by the pair $(/, \{\text{numerator, denominator}\})$. When arities are natural numbers, terms are built from lists of subterms and so list-processing operations are involved. By considering arities to be sets, we use set operations instead. These operations can be described as categorical constructs in **FinSet**, constructs such as limits and colimits, and thus we develop a wholly categorical treatment.

We define a type of operators as below:

```
datatype opr = opr of symbol * (element Set)
```

Thus an operator is of the form `opr(s,a)` with `s` a symbol (its name) and `a` a set (its arity). Here `element` is the type of names of argument places. We may fix this to be, say, character strings or integers.

We define terms recursively as either a variable $\langle x \rangle$, denoted `var(x)`, or as an operator together with an assignment of terms to the argument places $\rho(t_{\alpha_1} \ldots t_{\alpha_n})$, denoted `apply(rho,fn x => t(x))`. The '|' is ML's labelled union of these two cases.

```
datatype Term = var of element |
                apply of opr * (element -> Term)
```

A term substitution $f : a \to b$ is a function taking elements of set a to terms whose variables are in set b. We also need to keep track of the source set a and the target set b, so we define a substitution to be a triple:

```
datatype Substitution =
subst of (element Set)*(element -> Term)*(element Set)
```

Following the treatment in Section 3.2.7, composition of substitutions is defined in terms of the application of a substitution to a term.

```
fun subst_apply(t)(S as subst(a,f,b)) =
    case t of var(x) => f(x)
            | apply(psi,s) =>
              apply(psi,fn x => subst_apply(s x)(S))
```

```
fun subst_compose(S as subst(c,g,d),subst(a,f,b)) =
    if seteq(b,c)
       then subst(a,fn x => subst_apply(f x)(S),d)
       else raise non_composable
```

We also need the identity substitution on a set, together with source and target functions, which are defined as follows:

```
fun subst_ident(a) = subst(a,fn x => var(x),a)
fun subst_s(subst(a,_,_)) = a
fun subst_t(subst(_,_,b)) = b
```

Define the category $T_\Omega{}^{Fin}$ as the 4-tuple of these functions:

```
val FinKleisli =
    cat(subst_s,subst_t,subst_ident,subst_compose)
```

3.4.3 A finite category

As another example of a category in ML, consider a finite category, that of Section 3.2.3, depicted below.

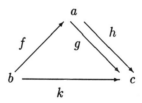

Composition is defined by $gf = k$ and $hf = k$ and the existence of an identity arrow for each object is assumed.

In this case the objects and arrows are constants. We represent them using enumerated types. Identities are constructed as labelled objects (we may make them additional constants but this lengthens the case analyses):

```
datatype Obj = a | b | c
datatype Arrow = f | g | h | k | id of Obj
```

Source, target and composition are defined explicitly for each argument using case analysis.

```
val example_category =
    let val s = fn f => b | g => a | h => a |
                    k => b | id(x) => x
        and t = fn f => a | g => c | h => c |
                    k => c | id(x) => x
        and ident = fn x => id(x)
        and comp =
          fn (id(x),u) => if t(u)=x then u
                           else raise non_composable
           | (u,id(x)) => if s(u)=x then u
                           else raise non_composable
           | (g,f) => k | (h,f) => k
           | _ => raise non_composable
    in cat(s,t,ident,comp) end
```

3.5   Functors

Functors are maps between categories that preserve the category struc-
ture. Let **A** and **B** be categories, a functor $F : \mathbf{A} \to \mathbf{B}$ consists of two
functions, one from the objects of category **A** to those of **B** and one from
the arrows of **A** to those of **B**. It is usual to denote both these functions
by the functor name, F. The sense in which the category structure is
preserved is given in the following definition.

Definition 3 *A functor $F : \mathbf{A} \to \mathbf{B}$ from category* **A** *to category* **B** *is a
pair of functions*

- $F : Obj(\mathbf{A}) \to Obj(\mathbf{B}), \quad F : a \mapsto F(a)$

- $F : Arrow(\mathbf{A}) \to Arrow(\mathbf{B}), \quad F : (f : a \to b) \mapsto F(f) : F(a) \to F(b)$

satisfying $F(i_a) = i_{F(a)}$ and $F(gf) = F(g)F(f)$ whenever gf is defined.

A functor from a category **C** to itself is called an *endofunctor* on **C**.
   Functors $F : \mathbf{A} \to \mathbf{B}$ and $G : \mathbf{B} \to \mathbf{C}$ may be composed to form
a functor $GF : \mathbf{A} \to \mathbf{C}$ by the separate composition of the constituent
functions. Identity functors consists of two identity functions. In fact,
we may consider categories to be objects within a category and functors
to be arrows.

3.5.1   Functors computationally

Functors, consisting as they do of two functions, one on objects, the other on arrows, can be represented quite simply. We need only recall that functors, being arrows, determine their source and target categories, which therefore need to be included as components in the representation. Just as the type of a category depends on that of its objects and arrows, so the type of a functor depends upon the types of objects and arrows in both the source and target categories. We put all this together in a type definition for functors.

```
datatype ('oA,'aA,'oB,'aB)Functor =
      ffunctor of ('oA,'aA)Cat *
                  ('oA->'oB) * ('aA->'aB) *
                  ('oB,'aB)Cat
```

The word 'functor' is rather overused, occurring not only in category theory but also as programming constructs in languages like Prolog and ML (the latter is related to the categorical notion of functors). Note that we write ffunctor rather than functor to distinguish categorical functors from ML functors.

3.5.2   Examples

To describe particular functors we need the source and target categories and the functions on objects and on arrows.

Let us consider an example, the identity functor on a category, $I_{\mathbf{A}}$: $\mathbf{A} \to \mathbf{A}$, acting identically on objects and arrows:

```
fun I(A) = ffunctor(A,fn x => x,fn x => x,A)
```

For a more substantial example consider the functor $\mathcal{X} : \mathbf{FinSet} \to \mathbf{FinSet}$ which maps a finite set a to $a \times a$, the cartesian product of a with itself. On arrows (typed functions) it maps f to $f \times f$ defined by $(f \times f)(x,y) = (f(x), f(y))$. The functorial nature of $\mathcal{X}$ can be established as follows:

$$\mathcal{X}(gf)(x,y) = (gf(x), gf(y)) = (g(f(x)), g(f(y)))$$
$$= \mathcal{X}(g)(f(x), f(y)) = \mathcal{X}(g)\mathcal{X}(f)(x,y)$$

and

$$\mathcal{X}(i_a)(x,y) = (x,y) = i_{\mathcal{X}(a)}(x,y)$$

We now program this functor. The cartesian product of finite sets may be defined recursively as follows:

```
fun cartesian_prod(A,B) =
        if is_empty(A) then nil_set else
          let val (a,A') = singleton_split A in
          mapset(fn b => (a,b),B)  U  A' X B end
```

Here `mapset(f,C)` is the image of set `C` through the function `f` and `U` is set union.

The product of two set arrows is defined below:

```
fun prod_arrow(set_arrow(A,f,B),set_arrow(C,g,D))  =
        set_arrow( cartesian_prod(A,C),
                   (fn (x,y) => (f(x),g(y))),
                   cartesian_prod(B,D) )
```

The functor $\mathcal{X}$ is the 4-tuple consisting of the source category, the two functions, one on objects, the other on arrows, and the target category:

```
val X =
  ffunctor(FinSet,fn A => cartesian_prod(A,A),
                  fn f => prod_arrow(f,f),FinSet)
```

Category theory leads us to ask not only what structures are involved in a computation but also what are the appropriate maps between structures. Functors are a computation on structures together with a computation on the maps between structures. These two functions are connected by the functorial axioms. This should be contrasted with collections of loosely connected or totally unconnected functions which are grouped together to form *ad hoc* modules. Not only is there a mathematical rationale to this collection of functions but it provides a more complete description of the situation. The completeness of the description is important in ensuring a wider re-usability of the code.

3.6 Duality

Like several other branches of mathematics, category theory supports a principle of duality. In the case of category theory, this arises from the invariance of the theory under the interchange of the source and target of arrows, i.e. reversing arrows. Each categorical property has a dual obtained by reversing all arrows in the statement of the property. The invariance means that one proof gives two results, the theorem proven and its dual. In programming, one piece of code will perform two calculations!

Duality can be formalized at a meta-level as a transformation of formulae in the first-order theory of categories. Each formulae has a dual obtained by a purely syntactic transformation interchanging sources and targets and reversing composites, see [Mac Lane 71] for details. This is interesting as a theorem-proving technique in which a theory is mapped back into itself under a validity-preserving map. In programming this syntactic manipulation of code goes under the name of 'program transformation' and usually arises in the context of efficiency of execution rather than multiple computations.

As pointed out by Mac Lane [1971], there is an alternative description of duality involving operations on categories and functors. For a category **A**, we can define a 'dual' category, $dual(\mathbf{A})$, whose objects are those of **A** and whose arrows are the reverse of arrows in **A**. Composition has then to be reversed:

Definition 4 *Let* **A** *be a category. The* dual *category,* $dual(\mathbf{A})$, *has as objects those of* **A**. *Arrows* $f : a \to b$ *in* $dual(\mathbf{A})$ *are arrows* $f : b \to a$ *in* **A**. *Identities are those of* **A**, *whilst composition of* $f : a \to b$ *and* $g : b \to c$ *in* $dual(\mathbf{A})$ *is defined as* $fg : c \to a$ *in* **A**.
The dual *of a functor* $F : \mathbf{A} \to \mathbf{B}$, $dual(F) : dual(\mathbf{A}) \to dual(\mathbf{B})$, *is the same pair of functions as* F *but between the dual categories.*

Any categorical property or construction is transformed into its dual by interpreting it in the dual category. This is the role of duality in the forthcoming programming. The dual of a construction is obtained from a program for the construction by composition with duality operations. In Chapter 4 we use duality to transform colimit constructions to limit constructions and in Chapter 5 we deal with constructions of categories where duality introduces canonical isomorphisms.

Here all we do is code up the duality operation on categories and functors. This is a straightforward translation into ML of the above definitions:

```
fun dual_Cat(cat(s,t,i,c)) =
        cat(t,s,i,fn (g,f) => c(f,g))
fun dual_Fun(ffunctor(A,fo,fa,B)) =
        ffunctor(dual_Cat(A),fo,fa,dual_Cat(B))
```

3.7 An assessment*

We now assess what we have done in representing categories and functor
for computation, mentioning a few important points. A fuller treatment
will be found in Chapter 10, where we discuss formal (linguistic) aspects of category theory and the requirements on a language for encoding
category theory.

The representation of categories given above is intended to capture
exactly those ingredients of a category required in categorical constructions so that we may code up these constructions to run on computers. It
will become clear in the next few chapters that this representation indeed
allows us to program many basic constructions in category theory. The
generality of these constructions is captured through passing categories
as parameters, using the higher order types of ML, and through type
polymorphism.

Categories are expressed as types within ML. However, the constraint
provided by the type system is too weak. For the type correctness of a
program, any four functions taking arguments of the requisite type and
returning results of the requisite type can be considered to be a category.

Type systems for programming are a compromise between expressiveness and utility. ML polymorphism is more expressive than many of
the the type systems of current programming languages like Pascal, yet
is much weaker than systems designed for expressing constructive mathematics. Unlike many of the more expressive systems, the type-checking
of ML is static, i.e. it is decidable whether a value is of a particular type.
This property has advantages from a programming point of view, since
type security of a program can be automatically assured before running
the program.

The information which is lost in moving from the mathematics to ML
types is, first of all, the axioms relating the functions. For categories,
these are the axioms of associativity of composition, identities and those
defining the source and target of composites and identities. Moreover,
where partial functions occur, as in the composition of arrows in a category, there is no means of describing in ML types the domain of definition
of partial functions. Programming languages, designed as they are for
executability, separate the constructive elements in a proof from the verification that the construction satisfies its specification. It is, therefore,
incumbent upon the programmer to ensure that only structures satisfying
the requisite axioms are supplied to programs.

An alternative is to force this verification to take place. There has been a good deal of interest over the last few years in formal systems for mathematics which include proofs. Some of this interest has been generated by the requirements of computation. Not only do we wish to know the possibility of expressing mathematics in a formal system, but we also want it to be a practical proposition to run various parts of mathematics on a computer.

The computations we undertake involve finite structures. For example, we consider categories of finite sets and of finite graphs. This could be extended to enumerable structures such as lazy lists. However, there are categories where the objects are themselves described as types, such as categories of algebras and categories of categories. To express these in a type system we need type universes, at least a type of 'small' types.

Some further points about the categorical programming we describe in this book: We are not inventing a categorical programming language, like that of Hagino [1987], described later, or a language for programming category theory, except in as much as the type definitions and functions in ML constitute a language. Notice that the arrows in a category are not represented by the arrow type constructor for function types. The type constructors are associated with an underlying category whose objects are types and whose arrows are functions. Arrow types correspond to exponentials in the category. It is not this category that we are encoding. As a consequence, the well-formedness of composites is not a matter of type-checking. In fact, such well-formedness requires a computation of equality on objects in the category. In formalizing the notion of a category, this equality on objects is not usually made explicit but it has to be considered in any foundation for category theory, see [Bénabou 85]) for a discussion of this matter.

One method of handling equality on objects is to include an equality function as a component of the representation of a category. To insist that all categories have a computable equality on objects will rule out many categories in which the equality is not computable but which are otherwise perfectly well-behaved – categories which we can represent, compute with and usefully apply, for example, categories of functors. This suggests introducing two forms of categories – those with and without equality on objects. An analogous situation arises with equality on values in programming languages.

Notice also that, unlike the mathematics, we make an allowance for non-denoting expressions such as the composition of non-composable arrows. In cases like this, we include exception handling, where possible, to make the program 'robust', i.e. able to cope with erroneous input.

Some definitions of categories explicitly introduce the collection of arrows between each two objects a and b as a set called the hom-set and denoted, in category **C**, by $\text{Hom}_{\mathbf{C}}(a, b)$ or just $\mathbf{C}(a, b)$. We do not consider this part of the general structure of categories. Where we wish to consider hom-sets as finite sets, we represent them as an extra component of type 'o x 'o -> ('a Set) and hence define 'locally finite' categories.

So far, we have looked at the type of categories. Now let us look at a particular category, **FinSet**, to see how the computational representation matches the intended mathematics. In fact, the representation falls short of the category of all finite sets. The type of categories allows only 'homogeneous' categories – those in which all objects have the same type, and likewise all arrows. Thus the sets in FinSet all have elements of the same type. This is indicated by the fact that FinSet is a polymorphic constant of type ('a Set,'a Set_Arrow)Cat. Letting the type variable 'a be the type int, we see that FinSet may be the category of finite sets of integers. This uniformity of type in ML is intended to allow us to define types where computation is allowed. Thus in finite sets of integers, disjoint unions may be represented by particular arithmetic computations which would not be available in the general category of sets. By enriching the type structure, beyond that of ML, it is possible to encode more general categories of sets. For instance, dependent types allow us to express categories of homogeneous sets whose type of elements vary from set to set.

There is a further mismatch between the category **FinSet** and its representation. The arrows in **FinSet** are total functions yet the representation allows partial functions, since the arrow type constructor gives the space of all partial functions, allowing, as it does, for the non-termination of programs. Unless we are working within a framework where all functions are total, ensuring that functions are total on a domain is unlikely to be decidable. Again it is incumbent upon the programmer to ensure that total functions are supplied where required.

Some of the verification that structures are indeed categories can be avoided through systematic ways of constructing categories from other categories. Constructions of categories are introduced in Chapter 5 where we show that from a few primitive categories, like **FinSet**, a whole range of useful categories can be built and their categorical nature can be assured through the correctness of the construction.

3.8   Conclusion

In this chapter we have considered at some length the fundamental concepts of categories and functors, displaying types to represent these and writing programs for particular categories and functors. So far we have not presented any significant categorical constructions so the code we have written is fairly simple, all the effort going into the design of representing types. In the next few chapters we consider constructive proofs of theorems in category theory. From these proofs we derive programs for calculations within categories, using the representations of categories and functors presented here.

3.9   Exercises

Exercise 1. Let us consider how to represent the category of finite graphs **FinGraph** in ML. This is a full subcategory of the category of graphs described in Section 3.2.2.

Objects in the category are finite graphs which may be represented as 4-tuples, consisting of two finite sets N and E (of nodes and edges) and two functions $s, t : E \to N$. We represent the collection of such graphs as a type in ML as follows:

```
datatype Graph =
         graph of (Node Set)*(Edge Set)*
                  (Edge->Node)*(Edge->Node)
```

The types `Node` and `Edge` are names for the nodes and edges and so may be fixed to be character strings or integers or whatever.

Arrows are pairs of functions,

$$(f : N \to N', g : E \to E') : (N, E, s, t) \to (N', E', s', t')$$

such that, for all $e \in E$, $f(s(e)) = s'(g(e))$ and $f(t(e)) = t'(g(e))$. This pair of functions, together with the source and target graphs form the type of arrows:

```
datatype Graph_Arrow =
graph_arrow of Graph*(Node->Node)*(Edge->Edge)*Graph
```

The source and target of arrows are projections onto the first and last components. Define ML functions for the identity arrows and the composition of arrows and hence define the category of finite graphs in ML.

Exercise 2. This is an exercise in defining the category $\mathbf{FinSet}_{Rel}$ (Section 3.2.1) in ML. The objects are finite sets and the arrows are relations. A relation may be represented as a finite set of pairs. Define arrows in $\mathbf{FinSet}_{Rel}$ as a type in ML (including the source and target sets).

Write code for the composition of relations and hence define the category in ML. For this you will need some operations on finite sets. Use the following (you may like to define them using the functions on sets described in the previous chapter): the mapset or image operation 'mapset(f,S)' applying a function f to all elements of a set S to yield the set $\{f(x) \mid x \in S\}$, the cartesian product $S \times T$ of two sets, the subset defined by a predicate P, 'filter S by $P = \{x \in S \mid P(x)\}$', and equality on sets assuming an equality on their elements.

An alternative representation of relations uses power sets. The power set $\mathcal{P}(a)$ of a set a is the set of all subsets of a. A relation $r : a \to b$ can be described as a partial function $r : a \to \mathcal{P}(b)$. Define the composition of relations in this representation and the category whose arrows are relations in this form.

We shall see later that the operations on relations used here, seemingly awkward to code, can be expressed elegantly using categorical constructions.

Exercise 3. Recall that a partial order can be considered to be a category (Section 3.2.5). Let us see how to represent the natural numbers $\mathcal{N}$ as a category under the usual partial order: $m \leq n$ iff $\exists p \in \mathcal{N} . m + p = n$.

Objects are natural numbers which, following Peano arithmetic, we may define to be a type whose values are in 1-1 correspondence with expressions built from a constant using one unary operation:

```
datatype Num = zero | succ of Num
```

Thus `succ(succ(zero))` represents the number 2. Operations like addition may be defined on natural numbers using recursion:

```
fun add(zero,n) = n
  | add(succ(m),n) = succ(add(m,n))
```

For each pair of natural numbers (m, n) such that $m \leq n$ there is to be exactly one arrow $m \to n$. For $m \leq n$, we represent the unique arrow $m \to n$ by the pair of numbers $(m, n - m)$. Moreover any pair of numbers (p, q) represents an arrow $p \to p + q$. Notice that the source and target numbers are reconstructable from the pair representing the order relation.

What are identity arrows? Define composition of arrows in ML and hence the category derived from the partial order on natural numbers.

Exercise 4. A monoid is a semigroup (a set with an associative binary operation) with an identity element. A category with just one object is a monoid, i.e. the arrows defining the category form a monoid whose binary operation is composition and, conversely, any monoid forms a category with one object.

The set of lists A^* of elements drawn from a set A form a monoid, with the binary operation as the end-to-end concatenation of lists and identity as the empty list. Define in ML the corresponding category.

Exercise 5. Define the relation $\preceq$ on lists by $s \preceq t$ iff s is an initial part of t, i.e. there is a list u such that $t = s \diamond u$ where $\diamond$ is the end-to-end concatenation of lists. An arrow between lists $f : s \to t$ is a function f from the elements of s such that $f(s) \preceq t$ where the application on the left is elementwise.

Define and the composition and identities and prove that the result is a category.

Choose a suitable representation in ML for the arrows (the analogy with Exercise 3 may be useful) and code up the category.

Exercise 6. Matrices Consider the category **Matrix**$_K$ whose objects are natural numbers and whose arrows $n \to m$ are $m \times n$ matrices over a commutative ring K. Composition is matrix product. Represent this in programming terms (you may wish to restrict attention to the case when K is the ring of integers).

Exercise 7. Consider the power set operation, mapping a set to its set of subsets $\mathcal{P} : a \mapsto \mathcal{P}(a)$. We show that it is functorial, that is, there is an accompanying operation on functions so as to form a functor. The operation on functions mapping $f : a \to b$ to $\mathcal{P}(f) : \mathcal{P}(a) \to \mathcal{P}(b)$ is defined by $\mathcal{P}(f)(s) = \{f(x) \mid x \in s\}$, the image of subset s under f. Verify that this defines a functor $\mathcal{P} : \mathbf{Set} \to \mathbf{Set}$ and program it for finite sets. The power set of a finite set may be computed recursively very much like the cartesian product of sets in Section 3.5.

Exercise 8. 2-categories

A richer compositional structure than a category is a 2-category, which has not only arrows but also arrows between arrows:

Definition 5 *A* 2-category *is a category* **A** *together with arrows between arrows with the same source and target. Objects in* **A** *are called 0-cells, arrows are called 1-cells and arrows between arrows are 2-cells. Let $f, g : a \to b$, denote a 2-cell by $\alpha : f \Rightarrow g$ or, more fully, $\alpha : f \Rightarrow g : a \to b$ or, pictorially, as:*

$$a \quad \xrightarrow{\quad f \quad} \Downarrow \alpha \xrightarrow{\quad g \quad} \quad b$$

There are two compositions of 2-cells: If $f, g, h : a \to b$ and $\alpha : f \Rightarrow g$, $\beta : g \Rightarrow h$, there is a composition $\beta . \alpha : f \Rightarrow h$ forming a category for each pair of objects a and b. This is vertical composition. For 2-cells $\alpha : f \Rightarrow f' : a \to b$ and $\beta : g \Rightarrow g' : b \to c$ there is an associative composition $\beta \circ \alpha : gf \Rightarrow g'f'$. This is horizontal composition, often written as $\beta\alpha$. These two compositions satisfy the so-called interchange law:

$$(\alpha . \alpha') \circ (\beta . \beta') = (\alpha \circ \beta) . (\alpha' \circ \beta')$$

Finally, identities of vertical composition are those of horizontal composition, in the sense that if $i_a : a \to a$ is the identity on a and $i_{i_a} : i_a \Rightarrow i_a$ is the identity of '.', then it is also the identity of 'o'.

Following the representation of categories, represent 2-categories as an ML type.

Exercise* 9. Categories form objects in a category **Cat** with functors as arrows. Show this and consider how to encode **Cat** in a formalism in which type universes are available.

Chapter 4

Limits and Colimits

Category theory is the theory of typed composition of arrows and as such is a very 'weak' theory. The theory finds its strength in powerful and intricate descriptive mechanisms. Amongst these are the topic of this chapter – limits and colimits.

Limits and colimits give a unified treatment of constructs such as products and sums of pairs of objects and limits of chains of objects. These constructs appear in various guises in set theory, algebra and topology. For example, we may form the product of two sets, the product of two groups and the product of two topological spaces. The details of the construction vary, but the general form of products is common to all cases. Mac Lane [1948,50] showed that products can be described abstractly in terms of the projection arrows from products to their components. This description uses universality – a method of definition by which a class of structures is defined and a particular structure isolated by its relationship with others in the class. We look at this in detail below. Kan [1958] extended Mac Lane's observation, introducing a general concept of limit and colimit which encompasses many of the universally defined combinations of structures.

As general mechanisms for combining structures, limits and colimits are of relevance to programming. This observation goes back to the work of Goguen [1971] and Goguen and Ginali [1978] in general systems theory. Colimits, in particular, allow us to build structured objects from components. Notice that, unlike sets, objects with internal structure, like graphs, cannot in general be decomposed as disjoint unions. Decompositions have to take account of shared sub-components. Colimits handle this by introducing arrows between components of a structure.

In programming, as in any activity involving writing definitions, we give names to values, so creating environments. Environments are naming structures and may be sets of names, maps, labelled graphs, program modules, or algebraic theories, depending upon context. Building environments from components involves some, often fairly intricate, manipulation of names (variables or identifiers). We create names uniquely, identify them, separate them, or rename them as required. Limits and colimits provide general mechanisms for handling names. Coproducts (a form of colimit) correspond to separating names and often replace the requirement to create new variables distinct from those present (as with the non-functional 'gensym' of Lisp). On the other hand, pushouts (another form of colimit) correspond to identifying selected names. We look at examples of this later, particularly in Chapter 9 where we consider constructing algebraic theories.

In this chapter we present the first substantial programming exercises. The programs are drawn from constructive proofs in category theory. The correspondence between constructive proofs and programs allows proof techniques to be used in program construction. This has been influential in the development of type theory, as in [Martin Löf 75,82,84]. The Nuprl system [Constable et al. 85] is an implementation of a type theory in which programs can be automatically extracted from constructive proofs. The programming we undertake relies on the fact that category theory is largely (though not wholly) constructive: proofs of existence are explicit constructions which may be expressed as programs.

The programs derived from category theory are tightly organized, following the organization of the theory itself. They are also highly abstract being parameterized over categories. This gives an abstraction beyond that normally encountered in programming. Programs are 'data independent' in the sense that, by supplying different categories, programs can be executed on different data structures. A construction can be executed on sets, graphs, algebras or whatever depending upon the parameter category.

The inherent duality of category theory, its invariance under 'reversing arrows', is here put to computational use. Limits are dual to colimits, and because it is duality, colimits are dual to limits. Not only does duality allow us to interchange between the two concepts, and hence their representations as types, but also programs constructing colimits can be re-used to construct limits, simply by interpreting the colimit computation in the dual category.

A remarkable aspect of category theory is its richness in ways of constructing colimits from other colimits and, by duality, limits from other limits. These constructions have an easy pictorial nature, often being expressed as the 'pasting together' of diagrams. We use these constructions to capture the compositional nature of various algorithms, including, as we shall see later, algorithms for the unification of terms. As well as constructions obtained from simple diagrams, there are general constructions for computing arbitrary colimits in a category. The constructions are not in any sense 'deep' but their variety is fascinating.

We shall look in detail at some of these constructions. The proofs are based upon 'diagram-chasing'. When converted into ML programs, their diagrammatic form is necessarily lost. To overcome this, a graphical input/output facility could be provided for the programs or, possibly, a more pictorial language used. An interesting aspect of constructions of limits and colimits is that the universal property is usually inherited, in the sense that the universality of the constructed object arises directly from the universality of its constituents. This imposes a tight structure on proofs making them almost automatic if the relevant diagrams fit together, a point of interest in proof mechanization.

In this chapter, we first discuss definition by universality. We then turn to colimits, introducing some special cases, then the general concept. We represent these as types in ML using a functional representation of universality. We then present some programs for computing colimits, including a calculation of arbitrary finite colimits in a category. This is reproduced for limits except that constructions of limits are obtained from those of colimits using duality. We consider colimits before limits, and in greater detail, as they appear more often as general constructional mechanisms. Colimits are applied to programming tasks in Chapters 8 and 9.

We concentrate in this chapter on computing colimits and limits in the category **FinSet**. In the next chapter, we show how the general constructions are used in other categories.

4.1   Definition by universality

Definition by universality is a familiar definition technique which, in category theory, provides the principle way of characterizing structure. We illustrate universal definitions with an example from set theory:

> The *union* of two sets, A and B, is the smallest set containing both A and B.

It consists of two parts:

- A definition of a class of objects which, in this case, is the class of all sets containing both A and B.

- A criterion for distinguishing a particular element in this class (or, more generally, a sub-class of elements) by its relationship to other elements. In the case above, the criterion is simply the 'smallest', i.e. contained within any other element in the class.

It turns out that category theory is a suitable formalism for expressing these universal definitions. Both the class of elements and the criterion for distinguishing a particular element are described in the same, arrow-theoretic, language. There is a standard format for such definitions in which the criterion for distinguishing a particular element is the unique existence of an arrow satisfying some given conditions. This defines objects only to within an isomorphism: if two objects both satisfy the requirements of the definition then they are isomorphic. Thus objects are viewed abstractly, that is independent of a particular representation. This characteristic of category theory is one sense in which the theory is 'abstract'. For objects defined to within an isomorphism, we shall adopt the usual convention of talking of 'the' object, when meaning any object in the isomorphism class.

Determining an object to within an isomorphism poses a problem from a computational viewpoint; which object in the isomorphism class is computed or is the whole class computed? Our treatment is algebraic in that all constructs appear as operations determining a unique object in the isomorphism class. Where isomorphisms are important, these are calculated as components of a structure and hence are themselves results of operations. If there is more than one way of performing a calculation, the results may differ by an isomorphism which itself is the result of a calculation.

Limits and colimits can both be defined through universality. More examples will be found in succeeding chapters. A curious result is the interdefinability of universal concepts in category theory. Universal concepts can be obtained from one another by interpreting them in suitable categories.

4.2   Finite colimits

In this section we consider some special cases of colimits: initial objects, binary coproducts, coequalizers and pushouts. We define these concepts,

show how they may be expressed as types in ML, including a representation of the universality, and then encode examples in **FinSet**.

4.2.1 Initial objects

The definition of an initial object in a category goes as follows:

Definition 6 *An* initial object *in a category is an object, a, such that
for any object, b, there is a unique arrow* $f : a \to b$.

In the categories **Set** and **FinSet**, the initial object is the empty set
ϕ, because for any set b there is just one function from ϕ to b, that with
the empty graph.

We show how initial objects may be represented as a type in ML. An
initial object is an object, a, which satisfies a certain universal property.
This property defines a function taking each object, b, into the unique
arrow $f : a \to b$. It is this pair of an object and a function (assigning an
arrow to each object) which serves as a representation of an initial object.
If objects have type 'o and arrows have type 'a then initial objects have
type:

```
'o * ('o -> 'a)
```

We thus introduce the following type declaration to name types of the
above form:

```
type ('o,'a)InitialObj = 'o * ('o -> 'a)
```

This functional representation of the universal property is crucial
to the programming of category theory. It is 'Skolemization', turning
∀∃-statements (unique existence in this case) into functions. It applies
not only to the initial object but to all universally defined concepts in
category theory. We shall see many examples in the succeeding pages
of types of this form and show how to write programs to compute with
them. Notice that we could have used a **datatype** declaration here and
so introduced a constructor for initial objects. These definitions are
equivalent; we give the one above for simplicity.

The representation of initial objects as an ML type is fairly slack.
Conditions on the source and target of arrows, as well as the uniqueness of the universal arrow, are not included. However, it captures all
the structure necessary for computation. Further properties on the type
are treated as verification conditions as discussed in the previous chapter.

You may wonder whether it is necessary to include the universal property as a component of the type. It is often inherited through constructions and so universality of constructed objects can be guaranteed. However, the ability to handle universality as a structure in its own right makes various categorical constructions programmable – for example, free algebras and colimits in comma categories, both of which we meet later.

The initial object in **FinSet** is a value of the above type. We program it as follows:

```
val set_initial =
    (emptyset, fn a => set_arrow(emptyset,nil_fn,a))
```

Here `emptyset` is the empty set, which, if sets are represented as lists, is the empty list. The function `nil_fn` is the empty function that raises an exception on any argument.

In **Set** and **FinSet** there is but one initial object, the empty set. In other categories, initial objects may not be unique but are always isomorphic. We prove this using a simple but neat arrow-theoretic argument. The proof for other universal concepts is similar. Let a and a' be initial objects in a category. Since a is initial, there is an arrow $f : a \to a'$ and, since a' is initial there is an arrow $g : a' \to a$. Now $gf : a \to a$; but $i_a : a \to a$ is another arrow from a to itself, so by the uniqueness of the arrow from an initial object $gf = i_a$. Likewise, $fg = i_{a'}$. So a and a' are isomorphic.

We express this argument as the construction of an isomorphism, a pair of mutually inverse arrows, in the following little ML program. It takes two initial objects, including their universality, and returns an isomorphism:

```
fun iso_initial((a,univ_a),(a',univ_a')) =
                (univ_a(a'),univ_a'(a))
```

4.2.2 Binary coproducts

Binary coproducts are another example of colimits. They are defined as follows.

Definition 7 *A* (binary) coproduct *of objects a and b in a category is an object $a + b$ together with arrows $f : a \to a + b$ and $g : b \to a + b$ such that, for any object c' and arrows $f' : a \to c'$ and $g' : b \to c'$, there is a unique arrow $u : a + b \to c'$ (denoted $[f', g']$) such that the following diagram commutes.*

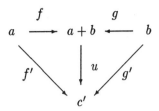

Binary coproducts determine a sum of arrows: Let $f : a \rightarrow a'$ and $g : b \rightarrow b'$, define $f + g : a + b \rightarrow a' + b'$ to be $[j_{a'}f, j_{b'}g]$ where $j_{a'} : a' \rightarrow a' + b'$, $(a' + b', j_{b'} : b' \rightarrow a' + b')$ is the coproduct of a' and b'.

The binary coproduct of sets a and b in **Set** (as well as in **FinSet**) is their disjoint union $a \uplus b$. The disjoint union can be represented in various (isomorphic) ways, labelling the elements of the two sets to keep them disjoint. There are functions $f : a \rightarrow a \uplus b$ and $g : a \rightarrow a \uplus b$, called the coproduct injection functions. The universal property takes a set c' and functions $f' : a \rightarrow c'$ and $g' : b \rightarrow c'$ and returns a function $u : a \uplus b \rightarrow c'$ defined by, for $x \in a \uplus b$ of the form $f(z)$ for $z \in a$ then $u(x) = f'(z)$, otherwise x is of the form $g(z)$ for $z \in b$ and $u(x) = g'(z)$. By definition, this function makes the following diagram commute.

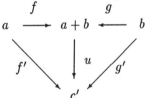

Moreover, any function u making this diagram commute must be defined as indicated on the image of f and on the image of g and hence u uniquely satisfies this commuting diagram.

We now turn to a representation of binary coproducts as a type in ML. A binary coproduct is a function which takes two objects a and b and returns a triple consisting of an object and two arrows, $(a + b, f, g)$. This triple satisfies a universal condition which defines a function from such triples to arrows, assigning to (c', f', g') the unique arrow $[f', g']$: $a + b \rightarrow c'$. As an ML type this becomes:

```
type ('o,'a)Coproduct =
        'o * 'o -> ('o,'a)Coproduct_CoCone
```

where `Coproduct_CoCone` is defined as:

```
type ('o,'a')Coproduct_CoCone =
        ('o * 'a * 'a) * ('o * 'a * 'a -> 'a)
```

Coproducts in **FinSet** are disjoint unions as explained above. We
show how they form a value of this type. To ensure the disjointness of
the union of two sets a and b, we label the elements of each set with a
label different from that of the other set. The two labels are arbitrary –
we choose pink and blue (for girls and boys). Recall that categories of
sets are homogeneous: all elements have the same type. Thus labelled
elements must have the same type as unlabelled elements. A new type
is introduced so that elements are closed under the labelling operation:

```
datatype 'a Tag =
   just of 'a | pink of ('a Tag) | blue of ('a Tag)
```

An example of a term of type int Tag is pink(pink(blue(just(2)))).

The coproduct object in **FinSet** of sets a and b is their (labelled) disjoint union. As an example, the disjoint union of set {just(1),just(2)}
with set {just(2),just(3)} is

```
{pink(just(1)),pink(just(2)),blue(just(2)),blue(just(3))}
```

The arrows into this coproduct are the labelling functions. The universal
property defines a function which takes any pair of functions $f : a \to c$
and $g : b \to c$ and yields a function $[f, g] : a + b \to c$.

All this is expressed in the following ML program:

```
fun set_coproduct(a,b) =
    let val sumab = mapset(pink)(a) U mapset(blue)(b)
            { the disjoint union }
        val univ =
          fn (c,set_arrow(_,f,_),set_arrow(_,g,_)) =>
                let val fg = fn pink x => f(x)
                              | blue x => g(x)
                in set_arrow(sumab,fg,c) end
            { the universal part }
    in (( sumab,
         set_arrow(a,pink,sumab),
         set_arrow(b,blue,sumab)),
       univ) end
```

4.2.3   Coequalizers and pushouts

We end this section with two further special cases of finite colimits: coequalizers and pushouts.

Definition 8 *A* coequalizer *of a parallel pair of arrows $f, g : a \to b$ in a category is an object c and an arrow $q : b \to c$ with $qf = qg$ such that for any object c' and arrow $q' : b \to c'$ such that $q'f = q'g$, there is a unique $u : c \to c'$ to make the diagram below commute.*

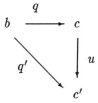

Coequalizers in **Set** and **FinSet** are calculated using equivalence classes. Let $f, g : a \to b$ be two functions. Define a relation on b by $x \rightsquigarrow y$ iff there is a $z \in a$ with $f(z) = x$ and $g(z) = y$. Let $\simeq$ be the equivalence closure of $\rightsquigarrow$, and c be the set of $\simeq$-equivalence classes. The quotient function $q : b \to c$ maps an element x to its equivalence class $[x]$ so that $qf = qg$. Now let c' be a set and $q' : b \to c'$ be a function such that $q'f = q'g$, define $u : c \to c'$ by $u([x]) = q'(x)$. This is well-defined and, moreover, $uq = q'$ by definition and u is clearly the only such function.

Definition 9 *A* pushout *of a pair of arrows $f : a \to b$ and $g : a \to c$ in a category is a pair of arrows $p : b \to d$ and $q : c \to d$ such that the square*

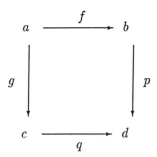

commutes and the following universal condition is satisfied: Suppose $p' : b \to d'$ and $q' : c \to d'$ with $p'f = q'g$, then there is a unique $u : d \to d'$ such that the following diagram commutes.

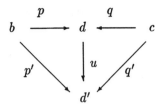

In **Set** and in **FinSet**, pushouts are constructed as a combination of a disjoint union and a quotient. Let $f : a \to b$ and $g : a \to c$ be a pair of functions and $b \uplus c$ the disjoint union of b and c with injection functions $j_b : b \to b \uplus c$ and $j_c : c \to b \uplus c$. Define the relation $\leadsto$ by $x \leadsto y$ iff there is a $z \in a$ with $j_b f(z) = x$ and $j_c g(z) = y$. Let d be the set of equivalence classes of $b \uplus c$ under the equivalence closure of $\leadsto$. Then d is the pushout object. We leave it as an exercise for the reader to prove this.

Functional representations in ML of these coequalizers and pushouts follow the pattern of initial objects and coproducts above:

```
type ('o,'a)Coequalizer =
      'a * 'a -> ('o * 'a) * ('o * 'a -> 'a)
type ('o,'a)Pushout =
      'a * 'a -> ('a * 'a) * ('a * 'a -> 'a)
```

Programs for coequalizers and pushouts in **FinSet** involve fairly intricate recursions. We leave this until later when we can make use of general constructions to calculate these colimits.

Exercise 1. Complete the construction of pushouts in **Set** and show that this indeed constructs a pushout.

Despite the caveat above, you might try to write programs which calculate, from first principles, coequalizers and pushouts in **FinSet**.

4.3   Computing colimits

We now turn to constructions of colimits showing how programs are derived from categorical proofs. From the wide variety available we choose several fairly simple constructions to illustrate the techniques involved. We then present a program for constructing arbitrary finite colimits.

We start with a construction of ternary coproducts from binary coproducts. A ternary coproduct is a coproduct of three objects and is defined like the binary coproduct, only replacing two objects by three.

Proposition 1 *A category with binary coproducts has ternary coprod-ucts.*

The proof is a typical diagram-chase establishing the inheritance of universality:

Given objects a, b and c in a category which has binary coproducts we can form the coproduct of these three objects as in the following diagram:

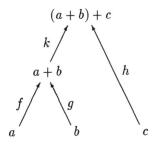

Before looking at universality, let us encode the construction in this picture.

Suppose we are given a composition function comp in a category and binary coproducts in it, that is a function b_coprod taking two objects and giving as result the coproduct object, the two arrows into it and the universal part. The object $(a + b) + c$, and the three arrows to it, are given by the expression:

```
let val (sumab,f,g),univab = b_coprod(a,b)
    val (sumabc,k,h),univabc = b_coprod(sumab,c) in
(sumabc, comp(k,f), comp(k,g), h) end
```

We now show that this construction is universal. The universality of the ternary coproduct is inherited from that of the binary coproducts involved as follows: For any object d and arrows $f_a : a \to d$, $f_b : b \to d$ and $f_c : c \to d$, from the universality of the coproduct of a and b, there is a unique arrow $u : a + b \to d$ such that $uf = f_a$ and $ug = f_b$. Now we have arrows $u : a + b \to d$ and $f_c : c \to d$. Thus, using the universality of the coproduct of $a + b$ and c, there is a unique arrow $v : (a + b) + c \to d$ such that $vk = u$ and $vh = f_c$. This is pictured in the following diagram:

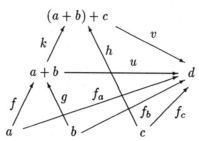

Thus v satisfies $v(kf) = f_a$, $v(kg) = f_b$ and $vh = f_c$. Moreover, the uniqueness is inherited: If v' satisfies these equations then, by the uniqueness of u, we have $v'k = u$ and so, by the uniqueness of v, $v = v'$.

As a program, this universality is a function taking an object d and three arrows f_a, f_b and f_c and yielding the arrow v. It is defined as follows, using the universal parts univab of the coproduct $a + b$ and univabc of the coproduct $(a + b) + c$:

```
fn (d,fa,fb,fc) =>
    let val u = univab(d,fa,fb) in
univabc(d,u,fc) end
```

We combine these two pieces of code into a function extending binary coproducts to ternary coproducts. A ternary coproduct in a category with objects of type 'o and arrows of type 'a has type:

```
'o * 'o * 'o ->
    ('o * 'a * 'a * 'a) * ('o * 'a * 'a * 'a -> 'a)
```

The function is defined as follows:

```
fun ternary_coprod(C,b_coprod) =
        { arguments are a category and a coproduct }
    fn (a,b,c) =>
      let val ((sumab,f,g),univab)   = b_coprod(a,b)
          val ((sumabc,k,h),univabc) = b_coprod(sumab,c)
          val univ =
            fn (d,fa,fb,fc) =>
                let val u = univab(d,fa,fb)
              univabc(d,u,fc)  end  in
  ( (sumabc,compose(C)(k,f),compose(C)(k,g),h),
    univ)  end
```

Notice the functionality of this colimit computation. It takes as argument a category together with a binary coproduct in the category. It is in this sense that categorical code is 'data independent'. We may supply the category of finite sets to produce an operation on sets or, say, the category of finite graphs to produce an operation on graphs.

Exercise 2. The ternary coproduct can also be computed as $a + (b + c)$. Write the associated program. Moreover, there is an isomorphism between $a + (b + c)$ and $(a + b) + c$. Write a program which uses the universality to compute this isomorphism as a (mutually inverse) pair of arrows.

Exercise 3. We can do better than this – any category with an initial object and binary coproducts has all finite coproducts. Prove this and write the corresponding iterative program.

We consider a more interesting construction of colimits – the 'pasting-together' of pushout squares.

Proposition 2 *In the diagram below, if the two squares are pushouts, then so is the outer rectangle:*

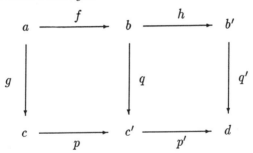

A pushout square in a category with arrows of type 'a consists of four arrows (the square) and a universal property (a function) and has type:

$$('a * 'a * 'a * 'a) * ('a * 'a -> 'a)$$

From two pushout squares of this type $((f,g,p,q),univfg)$ and $((h,q,p',q'),univhq)$ in a category C, we construct the new square as the outer rectangle, which is:

$$(compose(C)(h,f),g,compose(C)(p',p),q')$$

The universality of the outer rectangle is established as follows. Let $r : b' \to d'$ and $s : c \to d'$ such that $rhf = sg$. Then, since p, q is the pushout of f, g, there is a unique arrow $u : c' \to d'$ such that $uq = rh$ and $up = s$. Now, since $uq = rh$, by the universality of the second square, there is a unique $v : d \to d'$ with $vp' = u$ and $vq' = r$. Hence $vp'p = up = s$ and $vq' = r$. This is pictured below.

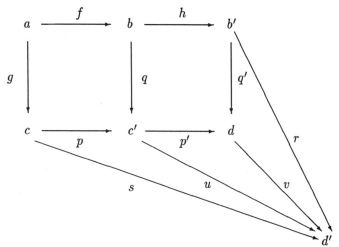

For the uniqueness of v satisfying these conditions, suppose that v' satisfies $v'p'p = s$ and $v'q' = r$. Then, by the uniqueness of u, $u = v'p'$, and so, by the uniqueness of v, $v = v'$.

Expressing this universality as a function we get the following:

```
fn (r,s) =>
    let val u = univfg(compose(C)(r,h),s) in
univhq(r,u)   end
```

Combining this with the pushout square above, we get a function to compose pushout squares:

```
fun compose_squares(C)( ((f,g,p,q),univfg),
                        ((h,q,p',q'),univhq) ) =
    ( (compose(C)(h,f),g,compose(C)(p',p),q'),
        fn (r,s) =>
            let val u = univfg(compose(C)(r,h),s) in
        univhq(r,u) end )
```

Exercise 4. Another operation on pushout squares is the coproduct. Prove that if (f, g, p, q) and (f', g', p', q') are pushout squares, then so is $(f + f', g + g', p + p', q + q')$. Write a program for the construction.

Exercise 5. Textbooks abound in constructions of colimits based upon combining diagrams. Here is another:

Proposition 3 *A category with binary coproducts and coequalizers has pushouts.*

The pushout of $f : a \to b$ and $g : a \to c$ is constructed as the coequalizer of $j_b f$ and $j_c g$ where j_b and j_c is the coproduct of b and c. Prove this and program up the construction. It can be used to compute pushouts in **FinSet**.

4.4 Graphs, diagrams and colimits

Colimits are a generalization of the universally defined concepts that we have met so far. The generalization is based upon diagrams in a category. We define diagrams and then the general concept of a colimit. We show how these are represented as types in ML and then give a construction of arbitrary finite colimits in a category.

Recall, from the previous chapter, the definitions of graphs and diagrams:

Definition 10 *A graph is a pair N, E of sets (of nodes and edges) together with a pair of mappings $s, t : E \to N$ called source and target respectively. We write $f : a \to b$ when f is in E and $s(f) = a$ and $t(f) = b$. A finite graph is one in which N and E are finite sets.*

Definition 11 *A diagram in a category* **A** *is a graph (N, E, s, t) (its shape) and two functions $f : N \to Obj(\mathbf{A})$, $g : E \to Arrow(\mathbf{A})$ which respect sources and targets in the following sense: For each edge $e \in E$, $f(s(e)) = s_A(g(e))$ and $f(t(e)) = t_A(g(e))$, where s_A and t_A are source and target of arrows in* **A**.

For diagram Δ, we denote by Δ_n, the object at node n, and by Δ_e, the arrow at edge e.

Finite graphs and diagrams may be directly represented as types:

```
datatype Graph = graph of (Node Set)*(Edge Set)*
                           (Edge->Node)*(Edge->Node)
datatype ('o,'a)Diagram =
         diagram of Graph*(Node->'o)*(Edge->'a)
```

Here `Set` is the type constructor for finite sets and `Node` and `Edge` are names which may be strings of characters or integers, or whatever.

To link objects with diagrams, we introduce cocones (also called 'cones from a diagram'):

Definition 12 *A* cocone *in a category on a diagram* Δ *(the* base*) is an object* a *(the* apex*) together with, for each node* n *in* Δ*, an arrow* $\xi_n : \Delta_n \to a$ *such that for all edges* $e : m \to n$ *in* Δ *the following commutes.*

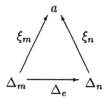

A cocone consists of three components, an apex, a base and a family of arrows indexed by the nodes in the diagram. All this is expressed as an ML type:

```
datatype ('o,'a)CoCone =
    cocone of 'o * ('o,'a)Diagram * (Node->'a)
```

In keeping with the categorical dictum of defining arrows as well as structures, we define arrows between cocones on the same base. A more general arrow which allows maps between bases is possible but is not needed here.

Definition 13 *A* cocone arrow *on diagram* Δ *from cocone* $\xi_n : \Delta_n \to a$ *on* Δ *to cocone* $\xi'_n : \Delta_n \to a'$ *is an arrow* $f : a \to a'$ *such that for all nodes* n *in* Δ *the triangle below commutes.*

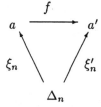

As a type in ML, this becomes:

```
datatype ('o,'a)CoCone_Arrow =
    cocone_arrow of ('o,'a)CoCone * 'a * ('o,'a)CoCone
```

Colimits are defined to be universal cocones in that there is a unique cocone arrow to any cocone on the same base:

Definition 14 *A* colimit *of a diagram* Δ *is a cocone, L, on* Δ *such that for any cocone on* Δ, *K, there is a unique cocone arrow* $u : L \to K$.

We represent this as an ML type:

```
type ('o,'a)Colimiting_CoCone =
  ('o,'a)CoCone * (('o,'a)CoCone -> ('o,'a)CoCone_Arrow)
type ('o,'a)Colimit =
                ('o,'a)Diagram -> ('o,'a)Colimiting_CoCone
```

The universality of colimits can be described by saying that colimits are initial cocones in the category of cocones. This is an example of the interdefinability of universal concepts in category theory and means that we could use the type of initial objects to provide an (equivalent) representation of colimits.

Definition 15 *A category having colimits of all (finite) diagrams is said to be* (finitely) cocomplete.

As a type, a cocomplete category is a category together with a colimit computing function:

```
datatype ('o,'a)CoComplete_Cat =
    cocomplete_cat of ('o,'a)Cat * ('o,'a)Colimit
```

Functors which preserve colimits are called cocontinuous – they preserve not just the apex of the colimiting cocone but the whole colimiting cocone.

Definition 16 *A functor* $F : \mathbf{A} \to \mathbf{B}$ *is* (finitely) cocontinuous *if, when* $\xi_n : \Delta_n \to a$ *is a colimiting cocone of a (finite) diagram* Δ *in* $\mathbf{A}$, *then* $F(\xi_n) : F(\Delta_n) \to F(a)$ *is a colimiting cocone of* $F(\Delta)$ *in* $\mathbf{B}$.

As a type, a cocontinuous functor is a functor together with a map between colimiting cocones:

```
datatype ('oA,'aA,'oB,'aB)CoContinuous_Functor =
    cocontinuous_functor of
            ('oA,'aA,'oB,'aB)Functor *
            (('oA,'aA)Colimiting_CoCone ->
                    ('oB,'aB)Colimiting_CoCone)
```

There is an alternative treatment of diagram as functors rather than as new structures. A diagram Δ in category $\mathbf{C}$ on graph A is equivalent to a functor $F : \mathbf{A} \to \mathbf{C}$ where $\mathbf{A}$ is the transitive, reflexive closure of graph A (called the path category of A). The correspondence is given by $F(a) = \Delta_a$ on nodes of A, and $F(f_1.f_2\ldots f_n) = \Delta_{f_n}\ldots\Delta_{f_2}\Delta_{f_1}$ on paths in A. Replacing diagrams by functors, means that fewer concepts need be considered as primitive in category theory and allows certain constructions to be concisely expressed, therefore giving compact code. However, the inclusion of the path category in the representation makes it rather inefficient in terms of storage. Ideally, program abstraction should make the choice immaterial for subsequent code. This requires an abstract description of diagrams and colimits. We consider the mathematics of this in the next chapter.

Exercise 6. The initial object is a special case of a colimit, being the colimit on the empty diagram. Prove this and write a program converting initial objects into colimits. Likewise, binary coproducts are colimits of two object diagrams, coequalizers are colimits of parallel pairs and pushouts are colimits of pushout diagrams (pairs of arrows with common source).

Exercise 7. Both identity functors and isomorphisms are cocontinuous. Isomorphisms are pairs of mutually inverse functors. Write expressions of the type of cocontinuous functors for these two cases.

Exercise 8. A functor $F : \mathbf{A} \to \mathbf{B}$ may be applied to a diagram Δ in $\mathbf{A}$ to yield a diagram $F(\Delta)$ in $\mathbf{B}$ by applying F to the objects and arrows of Δ. Likewise, functors may be applied to cocones. Write these two application functions.

4.5   A general construction of colimits

We have already seen how constructive proofs in category theory give rise to programs. We have also looked at some colimit constructions. In this section we present a theorem which guarantees finite cocompleteness of some categories. Its proof provides a function for calculating arbitrary finite colimits in a category.

The following theorem, which Manes [1976] describes as 'quite remarkable', is our starting point.

Theorem 1 *A category having an initial object, binary coproducts and coequalizers of parallel pairs of arrows has all finite colimits.*

This theorem asserts that if certain simple colimits exist in a category then so do colimits of arbitrarily complex finite diagrams. We look at the proof in detail and present the corresponding program.

Proof

The proof is in two parts. The first part constructs arbitrary finite coproducts (that is, colimits of discrete diagrams, those containing no edges). In the second part, we take account of edges in the diagram, using coequalizers.

Lemma 1 *A category* **C** *having an initial object and binary coproducts has all finite coproducts.*

This is proven by induction on the number of nodes in the diagram. If Δ is a discrete diagram in **C** then either:

1. Δ is empty, in which case the coproduct is the initial object; or

2. Δ is non-empty. Let n be a node in Δ and let Δ' be the diagram Δ without node n. Assume inductively that Δ' has a coproduct $\xi_j : \Delta_j \to a$ for $j \in nodes(\Delta')$. Let the binary coproduct of objects a and Δ_n be $f : a \to b$ and $g : \Delta_n \to b$. Construct a cocone on Δ, $\nu_j : \Delta_j \to b$ for $j \in nodes(\Delta)$, by $\nu_n = g$ and $\nu_j = f\xi_j$ otherwise. This is a coproduct of Δ, for suppose $\nu'_j : \Delta_j \to b'$ for $j \in nodes(\Delta)$ is a cocone on Δ then it is a cocone on Δ' and thus, by universality of ξ_j, there is a unique arrow $u : a \to b'$ such that for all $j \in nodes(\Delta')$ the diagram below commutes:

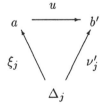

Now by universality of the binary coproduct, the two arrows $u : a \to b'$ and $\nu'_n : \Delta_n \to b'$ determine a unique arrow $v : b \to b'$ such that the diagram below commutes.

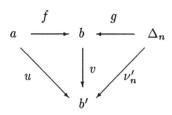

The two previous diagrams show that $v : b \to b'$ makes the diagram below commute for each $j \in nodes(\Delta)$. The uniqueness conditions of u and v ensure that v is the only such arrow:

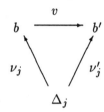

This proves Lemma 1.   □

Before progressing to the second part of the proof let us encode this construction of finite coproducts. To do this we need a category C and an initial object init and binary coproducts b_coprod in the category. The function derived from this proof then takes a discrete diagram (or, equivalently, a multiset of objects) and returns its colimit and hence has type:

```
finite_coproduct:
    ('o,'a)Cat * ('o,'a)Initial_Obj * ('o,'a)Coproduct ->
            (('o,'a)Diagram -> ('o,'a)Colimiting_CoCone)
```

In the case when the diagram in the category C is empty, we use the initial object init to construct a colimit of the empty diagram nil_diagram as in the following expression:

```
let val (i,i_univ) = init
    val i_cocone = cocone(i,nil_diagram,nil_fn)
        { initial object as apex and an empty base }
in (i_cocone,
    fn c1 =>
    cocone_arrow(i_cocone,i_univ(co_apex c1),c1)) end
```

For the case of the non-empty diagram D we follow the proof above using the binary coproduct function b_coprod and a recursive call to the function finite_coproduct:

```
let val (n,N1) = singleton_split(nodes(D))
    { first extract node n from the set of nodes }
    val (c,univc) =
        finite_coproduct(C,init,b_coprod)
                        (reduce_diagram(D,n))
    { compute the colimit of remaining diagram }
    val ((b,f,g),univcp) =
            b_coprod(co_apex(c),obj_at_node(D)(n))
    { this is the binary coproduct }
    val result_cocone =
        cocone(b,d,fn m => if m=n then g
                           else compose(C)(f,sides(c)(m)))
    { the universal part }
    val universal =
        fn c1 =>
        let val u = co_apex_arrow(univc(c1))
            val v =
                univcp(co_apex(c1),u,sides(c1)(n))
        in cocone_arrow(result_cocone,v,c1)   end
    in (result_cocone,universal)   end
```

Now for the second part of the proof – constructing finite colimits
from finite coproducts and coequalizers using coequalizers to account for
the edges in the diagram.

Lemma 2 *A category* **C** *having finite coproducts and coequalizers of parallel pairs of arrows has all finite colimits.*

Again proven by induction. If Δ is a finite diagram in **C** then either:

1. Δ is discrete (no edges) in which case the colimit is the coproduct;
   or

2. Δ has edges. Let $e : p \to q$ be an edge of Δ with associated arrow
   $f : \Delta_p \to \Delta_q$ and let Δ' be Δ without edge e. Assume inductively
   that Δ' has a colimit $\xi_n : \Delta_n \to a$. Consider now the parallel pair
   of arrows $\xi_p, \xi_q f : \Delta_p \to a$. Let the coequalizer be $h : a \to b$.

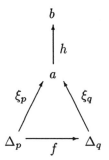

Construct a cocone on diagram Δ, $\nu_n : \Delta_n \to b$ for $n \in nodes(\Delta)$, by $\nu_n = h\xi_n$. We show it is colimiting. Let $\nu'_n : \Delta_n \to b'$ be a cocone on Δ. It is a cocone on Δ' and so, by the universality of ξ_n, there is a unique arrow $u : a \to b'$ such that for all $n \in nodes(\Delta)$, $u\xi_n = \nu'_n$ Now, $u\xi_p = u\xi_q f = \nu'_p$, hence, by the universality of the coequalizer, there is a unique arrow $v : b \to b'$ such that $vh = u$ as pictured:

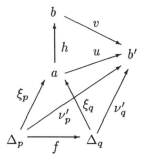

Thus v satisfies, for each $n \in nodes(\Delta)$, the commuting triangle below, where $\nu_n = h\xi_n$. It is the unique such arrow by the uniqueness in the definition of u and v.

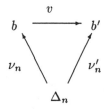

This completes the proof.　□

To program the second part of the proof, we start with the coequalizing operation associated with each edge in a diagram. This transforms a colimiting cocone (of the diagram minus the edge) to a colimiting cocone (of the augmented diagram). The program follows the steps of the proof and uses the same notation where possible.

```
add_edge : ('o,'a)Cat * ('o,'a)Coequalizer ->
              (('o,'a)Colimiting_CoCone * Edge ->
                              ('o,'a)Colimiting_CoCone)
fun add_edge(C,coeq) ((c,univ),e) =
   let val diagram(g,fo,fa) = base(c)
       val graph(N,E,s,t) = g
        { extracting base and its shape }
       val ((b,h),ce_univ) =
             coeq(sides(c)(s(e)),
                  compose(C)(sides(c)(t(e)),fa(e)))
        { the coequalizing }
        { now assemble the new colimiting cocone }
       val result_graph = graph(N,[e] U E,s,t)
       val result_diagram =
             diagram(result_graph,fo,fa)
       val result_cocone =
             cocone(b,
                    result_diagram,
                    fn n => compose(C)(h,sides(c)(n)))
        { the universal part }
       val universal =
             fn c1 =>
               let val u = co_apex_arrow(univ(c1))
                   val v = ce_univ(co_apex(c1),u)
               in cocone_arrow(result_cocone,v,c1) end
   in (result_cocone,universal) end
```

Combining these two parts, we arrive at a function which computes colimits of arbitrary finite diagrams:

```
finite_colimit : ('o,'a)IO_CP_CE_Cat -> ('o,'a)Colimit
```

It takes as argument a category together with the requisite colimits:

```
datatype ('o,'a)IO_CP_CE_Cat =
                io_cp_ce_cat of ('o,'a)Cat *
                                ('o,'a)InitialObj *
                                ('o,'a)Coproduct  *
                                ('o,'a)Coequalizer
```

Now for the definition of the function:

```
fun finite_colimit(cC as io_cp_ce(C,init,b_coprod,coeq))
              (d as diagram(graph(N,E,s,t),fo,fa)) =
    if is_empty(E)
      then finite_coproduct(C,init,b_coprod)(d)
      else let val (e,E1) = singleton_split(E)
              val d1 = diagram(graph(N,E1,s,t),fo,fa)
        in add_edge(C,coeq)((finite_colimit cC d1),e) end
```

Exercise 9. A category with coequalizers of parallel pairs of arrows has coequalizers of any finite set of parallel arrows. Prove this and write the iterative function.

Exercise 10. Mac Lane [1971 page 109] gives an alternative proof of the (dual of) finite cocompleteness of a category with initial object, binary coproducts and coequalizers. It uses the coequalizing operation only once. Look up this proof and encode it. Is it any more efficient?

4.6 Colimits in the category of finite sets

The category of finite sets **FinSet** is finitely cocomplete. This can be proven by observing that **FinSet** has initial object, binary coproducts and coequalizers and so, by Theorem 1, **FinSet** has all finite colimits. To compute these colimits we need only supply these special colimits in **FinSet** to the program derived from the proof of Theorem 1.

We have already programmed the initial object and binary coproducts in **FinSet**. Coequalizers are not quite as straightforward. We present a construction which is valid in any category and then use this as a basis of a recursive algorithm for coequalizers in **FinSet**.

Proposition 4 *In any category, if $q : b \to c$ is a coequalizer of the parallel pair:*

$$a \overset{f}{\underset{g}{\rightrightarrows}} b$$

and $r : c \to d$ is a coequalizer of:

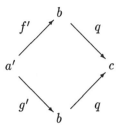

then the composite $rq : b \to d$ is a coequalizer of the following parallel pair:

$$a + a' \underset{[g,g']}{\overset{[f,f']}{\rightrightarrows}} b$$

This construction provides a recursive algorithm as follows: Decompose the source of the parallel pair non-trivially as a coproduct, recursively compute the coequalizers of the two parts then combine the results as indicated in the proposition. In **FinSet**, decomposition as non-trivial coproducts terminates.

To program coequalizers in **FinSet** we may either accumulate equivalence classes or choose representative elements from equivalence classes. We do the latter. The program itself is a case analysis: If the source set of a parallel pair is empty, the coequalizer is the identity on the target; if the source set is a singleton set, an explicit calculation is required, testing if the image elements are the same or not. The remaining case is the recursion.

```
fun set_coequalizer(set_arrow(s,f,b),set_arrow(_,g,_)) =
    { takes a parallel pair of set arrows }
  let val cat(_,_,id,comp) = FinSet in
    if is_empty(s) then
      { empty source - coeq is the identity }
      ((b, id(b)), fn (_,j) =>j)
    else if cardinality(s) = 1 then
      { singleton source - see if image elements
        are the same or different }
      let val (y,_) = singleton_split(s) in
        if f(y)=g(y)
        then ((b,id(b)),fn (_,j) => j)
        else
          let val b' = b diff (sing (g y))
              fun ff(z) = if z=g(y) then f(y) else z
```

```
              in ( (b',set_arrow(b,ff,b')),
                  fn (d',set_arrow(_,j,_)) =>
                          set_arrow(b',j,d') ) end end
        { the recursive case - use the proposition }
    else let val (a,a') = split(s)
             val ((_,q),univ) =
               set_coequalizer(set_arrow(a,f,b),
                               set_arrow(a,g,b))
             val ((d,r),univ') =
               set_coequalizer(comp(q,set_arrow(a',f,b)),
                               comp(q,set_arrow(a',g,b)))
         in ((d,comp(r,q)),
             fn (d',j)=> univ'(d',univ(d',j))) end end end
```

To compute finite colimits in **FinSet**, call the general colimit function
of the previous section supplying the category of finite sets, an initial
object in this category, binary coproducts (all coded previously) and
coequalizers (above).

```
val io_cp_ce_FinSet =
  io_cp_ce_cat(FinSet,set_initial,
               set_coproduct,set_coequalizer)
val cocomplete_FinSet =
  cocomplete_cat(FinSet,finite_colimit(io_cp_ce_FinSet))
```

Exercise 11. Prove Proposition 4 and encode the construction of co-
    equalizers in terms of others in an arbitrary category. Remember
    to include the universality of the construction.

4.7   A calculation of pushouts

We have produced some elaborate code but not yet shown it in action
calculating colimits. As an example, let us calculate the pushout in
FinSet of the following diagram.

$$\{A, B, C, D\}$$

$$f = \{A \mapsto C, B \mapsto C, \qquad g = \{A \mapsto B, B \mapsto E,$$
$$C \mapsto D, D \mapsto E\} \qquad\qquad C \mapsto B, D \mapsto F\}$$
$$\{C, D, E, F, G\} \qquad \{B, E, F, G\}$$

We first encode this diagram in ML and then call the general colimit
routine specialized to the category **FinSet**.

A diagram is assembled from its underlying graph. For pushout diagrams this looks like:

```
val G = let val N = {word "a",word "b",word "c"}
            val E = {word "f",word "g"}
            fun src (word "f") = word "a"
              | src (word "g") = word "a"
            fun tgt (word "f") = word "b"
              | tgt (word "g") = word "c"
        in graph(N,E,src,tgt) end
```

A diagram is defined by assigning sets to nodes and functions (set arrows) to edges. The three sets are as follows.

```
val a_set = {just "A",just "B",just "C",just "D"}
val b_set = {just "C",just "D",just "E",just "F",just "G"}
val c_set = {just "B",just "E",just "F",just "G"}
```

The functions associated with edges f and g are defined below:

```
fun f_fn (just "A") = just "C"
  | f_fn (just "B") = just "C"
  | f_fn (just "C") = just "D"
  | f_fn (just "D") = just "E"
fun g_fn (just "A") = just "B"
  | g_fn (just "B") = just "E"
  | g_fn (just "C") = just "B"
  | g_fn (just "D") = just "F"
```

Then the pushout diagram is given by the following expression:

```
fun nodes_to_sets (word "a") = a_set
  | nodes_to_sets (word "b") = b_set
  | nodes_to_sets (word "c") = c_set
fun edges_to_arrows (word "f") =
          set_arrow(a_set,f_fn,b_set)
  | edges_to_arrows (word "g") =
          set_arrow(a_set,g_fn,c_set)
val po_diagram =
          diagram(G,nodes_to_sets,edges_to_arrows)
```

The pushout is obtained as the value of the following expression:

```
let val cocomplete_cat(_,set_colimit) =
                              cocomplete_FinSet
in set_colimit(po_diagram) end
```

It is a cocone with a universal part. The apex of this cocone is calculated to be the following set:

```
{ pink just "D", pink just "A", blue pink just "G",
        blue pink just "F", blue blue pink just "G" }
```

Let us consider how this result was arrived at. The general construction of colimits of arbitrary finite diagrams has been invoked. This calls upon initial object, coproduct and coequalizer of finite sets. Firstly, the coproduct of the sets at the nodes in the diagram is accumulated starting with the initial object (the empty set) and using the labels 'blue' and 'pink' to construct disjoint unions. Secondly, elements of the coproduct are removed by two coequalizer operations, one for each of the edges in the diagram.

The rest of the cocone has po_diagram as base and the sides are functions performing labelling and identification. The side from node **a** is a set arrow containing the following function:

```
fn just "A" => pink just "A" | just "B" => pink just "A"
 | just "C" => pink just "A" | just "D" => pink just "D"
```

The function associated with node **b** is:

```
fn just "C" => pink just "A"
 | just "D" => pink just "A"
 | just "E" => pink just "D"
 | just "F" => blue pink just "F"
 | just "G" => blue pink just "G"
```

Finally, that associated with node **c** is:

```
fn just "B" => pink just "A" | just "E" => pink just "A"
 | just "F" => pink just "D"
 | just "G" => blue blue pink just "G"
```

As well as all this being calculated, there is the universal part taking a cocone on po_diagram to a cocone arrow. It is left to the reader to work out what this function should be.

4.8 Duality and limits

Limits are dual to colimits – they are defined by reversing arrows in the definition of colimits. Dual to initial objects are terminal objects, to coproducts are products, to coequalizers are equalizers, and to pushouts are pullbacks. Dual to cocones are cones and to colimits are limits.

We showed in the previous chapter how duality can be programmed as an arrow-reversing operation on categories. We use this to calculate limits in categories by calculating colimits in the dual category. Thus the programs we have written are re-used to calculate limits.

In this section we introduce types for various limits, a duality operation converting colimits to limits, and a function for calculating limits of finite diagrams. In the next section we put this to use to calculate finite limits in **FinSet**.

Products, which are dual to coproducts, are defined by:

Definition 17 *A* (binary) product *of objects a and b in a category is an object $a \times b$ together with arrows (called* projection *arrows) $f : a \times b \to a$ and $g : a \times b \to b$ such that, for any object c' and arrows $f' : c' \to a$ and $g' : c' \to b$, there is a unique arrow $u : c' \to a \times b$ (denoted $\langle f', g' \rangle$) such that the diagram below commutes.*

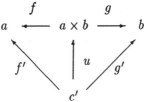

Just as the coproduct operation extends to arrows, we may, by duality, define the product $f \times g$ of arrows $f : a \to b$ and $g : a' \to b'$ as $\langle f\pi_a, g\pi_{a'} \rangle : a \times a' \to b \times b'$ where π_a, $\pi_{a'}$ are the product projections from $a \times a'$.

The representation of colimits in the previous section was phrased in terms of the objects and arrows involved but not including the direction of arrows, hence limits have the same type as the corresponding colimits. Thus the type of products is given by:

```
type ('o,'a)Product =
        'o*'o -> ('o*'a*'a) * ('o*'a*'a->'a)
```

The definitions of and type expressions for other limits are treated in the same way.

94   *LIMITS AND COLIMITS*

The general concept of a limit is treated similarly. For the sake of definiteness, we introduce new types with constructors to indicate that, for instance, cones differ from cocones even though the functional types in terms of objects and arrows are the same.

```
datatype ('o,'a)Cone =
    cone of 'o * ('o,'a)Diagram * (Node->'a)
datatype ('o,'a)Cone_Arrow =
    cone_arrow of ('o,'a)Cone * 'a * ('o,'a)Cone

type ('o,'a)Limiting_Cone =
      ('o,'a)Cone * (('o,'a)Cone->('o,'a)Cone_Arrow)
type ('o,'a)Limit =
      ('o,'a)Diagram -> ('o,'a)Limiting_Cone
datatype ('o,'a)Complete_Cat =
      complete_cat of ('o,'a)Cat * ('o,'a)Limit
```

The following simple argument shows how colimit constructions convert to limit constructions. The dual of a diagram is given by reversing the edges of the underlying graph, giving a diagram in the dual category:

```
fun dual_graph(graph(N,E,s,t)) = graph(N,E,t,s)
fun dual_diagram(diagram(g,nm,em)) =
                  diagram(dual_graph(g),nm,em)
```

A colimiting cocone in a category $dual(\mathbf{C})$ on diagram $dual(\Delta)$ is a limiting cone in $\mathbf{C}$ on Δ. Thus there is a function taking colimiting cocones to limiting cones, defined as follows:

```
fun dual_cone(cocone(a,D,f),univ) =
    let val result_cone = cone(a,dual_diagram(D),f)
        val universal =
          fn (c1 as cone(a1,D1,f1)) =>
            let val c2 = cocone(a1,dual_Diag D1,f1)
            in cone_arrow( c1,
                           co_apex_arrow(univ c2),
                           result_cone) end
    in (result_cone,universal) end
```

Suppose that F is a colimit, i.e. a function taking diagrams to colimiting cocones, then, by conjugation with the two functions above, we have the corresponding limit as follows:

```
fun dual_colimit(F) =
    fn D => dual_cone(F(dual_diagram D))
```

This then is the function which converts colimit calculations into limit calculations.

We now return to Theorem 1 which asserted the cocompleteness of certain categories. The dual says:

Theorem 2 *A category having a terminal object, binary products and equalizers of parallel pairs of arrows has all finite limits.*

The proof is an appeal to duality. To use it to compute limits in category **C**, proceed as follows. Calculate the terminal object, binary products and equalizers of parallel pairs of arrows in **C**. These correspond, in the dual category, to the initial object, binary coproducts and coequalizers. Apply the general colimit extension function `finite_colimit` to calculate finite colimits in the dual category. Now use the above function `dual_colimit` to convert to limits in the original category. All this is captured in the following expression which calculates finite limits in a category:

```
dual_colimit(finite_colimit(io_cp_ce_cat(dual(C),te,pr,eq)))
```

Exercise 12. Show that in **FinSet** the terminal object is a one element set, products are cartesian products and equalizers are a subset construction.

4.9   Limits in the category of finite sets

As explained above we use duality to compute limits. We apply this to limits in **FinSet**. Firstly, we encode a terminal object (a one element set), binary products (the cartesian product) and equalizers (a subset construction). These are straightforward to encode as long as we ensure that the type of elements is closed under the requisite operations: pairing for products and a constant element (which we call `ttrue`) for the terminal object. We extend the type of tagged values to include these two operations:

```
datatype 'a Tag = ... | pair of ('a Tag)*('a Tag) | ttrue
```

Below are programs for these three limits in the category of finite sets.

```
val set_terminal =
     let val t = singleton_set(ttrue)
          { the terminal object is a one element set }
     in (t, fn a => set_arrow(a,fn x => ttrue,t)) end
          { the universal part is a constant function }

fun set_product(a,b) =
  let val a_cross_b = mapset(pair)(a X b)
          { the labelled cartesian product }
      val proj_a =
        set_arrow(a_cross_b,fn pair(y,z) => y ,a)
      val proj_b =
        set_arrow(a_cross_b,fn pair(y,z) => z ,b)
          { projections to first and second components }
      val univ =
        fn (p,f1,f2) =>
          set_arrow( p,
                     fn y => pair(f1 OF y,f2 OF y),
                     a_cross_b )
          { universal part - pairing two functions }
  in ((a_cross_b,proj_a,proj_b),univ) end

fun set_equalizer(f,g) =
  let val a = source(FinSet)(f)
      val e = a filtered_by (fn y => (f OF y)=(g OF y))
          { the equalizer object-a subset of the source }
  in ((e,set_arrow(e,fn x => x,a)),
          { the equalizer object and an inclusion }
     (fn (e1,h1)=>set_arrow(e1,fn y => h1 OF y,e))) end
          { the universal part }
```

We extend these special limits to limits of arbitrary finite diagrams in **FinSet** using duality:

```
val complete_FinSet =
     complete_cat( FinSet,
                   dual_colimit(finite_colimit(
                        io_cp_ce_cat( dual(FinSet),
                                     set_terminal,
                                     set_product,
                                     set_equalizer))))
```

4.10 An application: operations on relations

In this section we show how to implement operations on relations using limits and colimits.

A relation between elements in a set A and those in a set B is a subset of $A \times B$. There is a categorical notion of relations which is defined in any category and allows us to express operations for combining relations as limits and colimits.

A *span* in a category $\mathbf{C}$ is an ordered pair of arrows with common source:

A relation determines a span in **Set**: the projection functions from a set of pairs is a span. In general, spans in **Set** correspond to multirelations. Multirelations are bipartite multigraphs – relations in which there may be more than one witness in r to a pair in $a \times b$. If the sets a and b in the span are the same, then the span is a graph – a directed multigraph.

Let $\mathbf{C}$ be a category with pullbacks. Consider two relations in $\mathbf{C}$ with common intermediate object $(f : r \to a, g : r \to b)$ and $(f' : r' \to b, g' : r' \to c)$. We may form the pullback p, p' of g with f' as depicted below:

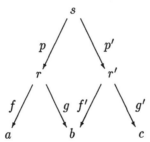

The *join* of the relations (f, g) with (f', g'), as introduced by Codd [1970], is the ternary relation $(fp, gp, g'p')$. The composition $(f', g') \circ (f, g)$ is the binary relation $(fp, g'p')$. As an exercise, show that this corresponds, in the category **Set**, to the standard composition of relations defined as follows: The composition of relation $r \subseteq a \times b$ with $s \subseteq b \times c$ is the relation $s \circ r$ defined by

$$(x, z) \in s \circ r \quad \text{iff} \quad \exists y : (x, y) \in r \text{ and } (y, z) \in s.$$

Before seeing how the pullback definition of composition yields a program, let us have a look at a straightforward program in which we represent relations as sets of pairs and use the above definition of composition. We use the language SETL [Dewar 78] to express this as it has a construct for iterating over sets:

```
PROGRAM composition
    sr := NIL
    LOOP (FORALL (x,z) IN axc) DO
        LOOP (FORALL y IN b) DO
            IF ((x,y) IN r) AND ((y,z) IN s) THEN
                sr := sr WITH (x,z)
            END IF
        END LOOP
    END LOOP
END PROGRAM composition
```

This program directly implements the definition of composition. It is a nested pair of loops. Efficiency can be improved by exiting from the inner loop once a suitable y is found.

The description of the composition in terms of a pullback gives the following program in any finitely complete category.

```
fun relation_compose(cC)((f',g'),(f,g)) =
    let val complete_cat(C,_) = cC
        val (p,p'),_ = pullback(cC)(g,f') in
    (compose(C)(f,p),compose(C)(g',p')) end
```

Notice that the program is linear – no repetitive constructs – we use the recursion inherent in the pullback in the category cC. This is an example of categorical primitives encapsulating iteration and recursion. Notice also that, because it is parameterized over categories, there is an implied generality to the code so that we may apply it to relations other than those in **FinSet**. For instance, we may wish to impose structure on the sets, and have relations preserving this structure.

To illustrate what is involved in a limit computation, let us run an example in the category **FinSet**. We have already encoded finite limits in **FinSet**. To make the description of relations as spans more succinct, we introduce the conversion function from sets of pairs:

```
fun relation_span(a,r,b) =
      { r is a subset of a x b }
      (set_arrow(r,fn pair(x,y) => x,a),
       set_arrow(r,fn pair(x,y) => y,b))
```

Define two relations, r from a to b and s from b to c:

```
val a = {just(1),just(2)}
val b = {just(3),just(4),just(5)}
val c = {just(6),just(7)}

val r = relation_span( a,
                        {pair(just(1),just(3)),
                         pair(just(2),just(3)),
                         pair(just(2),just(4))},
                        b )
val s = relation_span( b,
                        {pair(just(3),just(6)),
                         pair(just(4),just(6)),
                         pair(just(4),just(7)),
                         pair(just(5),just(6))},
                         c )
```

As a set of pairs, the composition of r with s is $\{(1,6),(2,6),(2,7)\}$.
As a span, the composition of the two relations is given by the expression

```
relation_compose(complete_FinSet)(r,s)
```

which has the following value:

```
let val r_o_s =
 {pair(pair(just(1),just(3)),
      pair(just(3),pair(pair(just(3),just(6)),ttrue))),
  pair(pair(just(2),just(3)),
      pair(just(3),pair(pair(just(3),just(6)),ttrue))),
  pair(pair(just(2),just(4)),
      pair(just(4),pair(pair(just(4),just(6)),ttrue))),
  pair(pair(just(2),just(4)),
      pair(just(4),pair(pair(just(4),just(7)),ttrue)))} in
( set_arrow( r_o_s,<function>,{just(1),just(2)} ),
  set_arrow( r_o_s,<function>,{just(6),just(7)} ) ) end
```

To understand how it got this result, recall that pullbacks are constructed from terminal objects, binary products and equalizers. We have encoded each of these for the category **FinSet**. The computation of the pullback uses duality and the colimit extension program (from Theorem 1). It is computed as a product iterated over the nodes, beginning with the terminal object {ttrue}. This explains the depth of nesting of pairs and the presence of ttrue in the result. Following the iterated product, an equalizer is computed for each of the two edges in the pullback diagram. This extracts the relevant subset of the product. Notice that there are four elements in the computed result where one would expect only three. The reason for this is that spans correspond to multirelations and there are two witnesses, the elements 3 and 4, to the pair $(2, 6)$. A factorization can be used to extract the relevant subset. The functions in the above result can be applied to arguments to check that they are indeed the correct projections.

Exercise 13. Intersection and union of relations Given two spans from a to b, (f, g) and (f', g'), in a category **C**, consider the limit $(u : s \rightarrow r, v : s \rightarrow r')$ of the diagram:

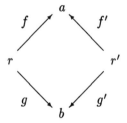

This determines a span (fu, gu) – the intersection of the two spans. Consider now the pushout of (u, v). Its universal property gives a span from a to b – the union of the two spans.

Work through the details, showing how these operations in **FinSet** correspond to the union and intersection of relations. Write the relevant programs.

4.11 Exercises

Exercise 14. In the previous chapter we defined a category $\mathbf{T}_\Omega{}^{Fin}$ whose arrows are term substitutions. Show that coproducts in $\mathbf{T}_\Omega{}^{Fin}$ are disjoint unions and express this as a program.

Exercise* **15.** The interchange of colimits theorem can be found in [Mac Lane 71], page 111. Write a program for this general colimit construction.

Chapter 5

Constructing Categories

So far we have managed to describe a few categories and to compute finite limits and colimits in the category of finite sets. What about other categories? Do we explicitly define each category that we are interested in and then, if needed, laboriously encode limits and colimits?

Fortunately not: there are systematic ways of constructing categories from other categories. When encoded as programs, these provide a means of introducing categories without explicitly defining them in terms of objects, arrows and associated operations. Moreover, under certain circumstances, limits and colimits in the constructed category are inherited from the constituent categories. In such cases, the inheritance itself is a construction which may be used to calculate limits and colimits in various categories.

Calculating limits and colimits is awkward for two reasons: the diagrams involved may be large and unwieldy and the objects and arrows may themselves be complex structures. In the previous chapter we gave an effective way of building large diagrams from small pieces whilst accumulating the colimit at the same time. In this chapter, we show how to separate the computation of colimits of complex objects into those of simpler objects.

We begin with comma categories, showing how this construction of categories is expressed as an ML program. We state and prove a theorem establishing cocompleteness of certain comma categories. This provides a program for computing colimits in these comma categories. We apply this to computing colimits of graphs. We then turn to categories whose objects are functors, showing how to compute colimits of diagrams of functors. Duality allows us to re-use these colimit programs for computing limits. This involves canonical isomorphisms linking duality and

103

constructions of categories. At the end of the chapter we present an abstract formulation of colimits – an abstract type for arbitrary colimits. This is prompted by the need for a unified treatment of representations of colimits and our quest for generality of code.

5.1   Comma categories

Comma categories were introduced by Lawvere [1963] in the context of the interdefinability of the universal concepts of category theory. The basic idea is the elevation of arrows of one category **C** to objects in another. The full generality can be obtained by taking a subclass of arrows – those whose source is in the image of one functor $L : \mathbf{A} \to \mathbf{C}$ and whose target is in the image of another functor $R : \mathbf{B} \to \mathbf{C}$.

Definition 18 *Let* $L : \mathbf{A} \to \mathbf{C}$ *and* $R : \mathbf{B} \to \mathbf{C}$ *be functors. The* comma category (L, R) *has, as objects, triples of the form* $(a, f : L(a) \to R(b), b)$ *where a is an object of* **A** *and b an object of* **B***. An arrow in* (L, R)*, from* (a, f, b) *to* (a', f', b') *is a pair of arrows* $s : a \to a'$ *and* $t : b \to b'$ *such that the following square commutes:*

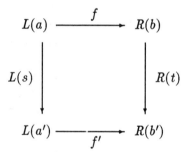

Composition is defined componentwise, $(s, t).(s', t') = (s.s', t.t')$ *and identities are pairs of identities.*

Associated with comma categories are two projection functors:

$$left : (L, R) \to \mathbf{A}, \qquad right : (L, R) \to \mathbf{B}$$

defined by $left(a, f, b) = a$ and $left(s, t) = s$ and similarly for the functor *right*.

Examples

An example of a comma category is the category of graphs. Graphs are finite, directed multigraphs, possibly with loops. Arrows of graphs are pairs of functions, mapping edges to edges and nodes to nodes, and preserving the source and target structure. This category is (isomorphic to) the comma category $(I_{\mathbf{FinSet}}, \mathcal{X})$ where $\mathcal{X} : \mathbf{FinSet} \to \mathbf{FinSet}$ is the functor $\mathcal{X}(a) = a \times a$ and, on arrows, $\mathcal{X}(f) = f \times f$ where $(f \times f)(x) = (f(x), f(x))$. An object of this comma category is a triple of the form $(E, f : E \to N \times N, N)$. Letting E be the set of edges, N the set of nodes and f the function allocating to each edge a pair of source and target nodes, we see that these are graphs. Moreover, arrows in $(I_{\mathbf{FinSet}}, \mathcal{X})$ are graph arrows.

Another example of a comma category is that of signatures (many-sorted operator domains). Categories of cones and of cocones are also comma categories. Comma categories arise through the operations of labelling (e.g. labelled trees), indexing and adding distinguished points.

5.1.1 Representing comma categories

We show how the comma category construction can be expressed as an ML program. Consider the comma category $(L : \mathbf{A} \to \mathbf{C}, R : \mathbf{B} \to \mathbf{C})$. To define this as an ML value, we need the type of objects and arrows in the category and the functions source, target, identity and composition. Let 'oA be the type of the objects in **A**, 'aA the type of its arrows, and similarly for the other categories. Objects in the comma category are triples of the source and target objects and the arrow in category **C**, and so have type: 'oA * 'aC * 'oB. Arrows in the comma category are pairs of arrows of type ('aA*'aB) together with the source and target objects:

```
datatype ('oA,'aA,'oC,'aC,'oB,'aB)Comma_Arrow  =
   comma_arrow of ('oA*'aC*'oB)*('aA*'aB)*('oA*'aC*'oB)
```

We define the comma category as a construction taking two functors and yielding a category:

```
fun comma_cat(L,R) =
  let val A = domain(L)
      val B = domain(R) in
cat( fn comma_arrow(Y,_,_) => Y,
     fn comma_arrow(_,_,Y) => Y,
     fn Z as (a,_,c) =>
        comma_arrow(Z,(identity(A)(a),identity(B)(c)),Z),
```

```
fn ( comma_arrow(_,(f,g),Z),
     comma_arrow(Y,(h,j),_) ) =>
 comma_arrow( Y,
               (compose(A)(f,h),compose(B)(g,j)),
               Z ) ) end
```

To represent a particular comma category we need only define the two functors involved. As an example, consider the category of graphs defined above. We need the functor $\mathcal{X} : a \mapsto a \times a$ encoded as follows:

```
val X =
    let fun prod(a,b) =
            mapset(pair)(cartesian_prod(a,b)) in
    ffunctor(FinSet,
             fn a => prod(a,a),
             fn set_arrow(a,f,b) =>
              set_arrow( prod(a,a),
                         fn pair(x,y)=>pair(f(x),f(y)),
                         prod(b,b) ),
             FinSet) end
```

The category of finite graphs is then the value of the following expression:

```
val cat_of_graphs = comma_cat(I(FinSet),cross_product)
```

Various special cases of comma categories can be defined. For instance, for a category $\mathbf{C}$ and object a of $\mathbf{C}$, we can form the comma category $(I_\mathbf{A}, K_a)$ where $K_a : \mathbf{1} \to \mathbf{C}$ is the constant functor from the one object, one arrow, category $\mathbf{1}$, returning object a of $\mathbf{A}$. Objects in this category are arrows into a. Arrows are commuting triangles. This is called a slice category and is denoted $\mathbf{C}/a$.

Exercise 1. An A-indexed set is a family of sets S_i, $i \in A$. Equivalently, it is a function $f : X \to A$. Show that these are equivalent (watch disjointness) and use the second representation to express the category of A-indexed sets as a comma category (a slice category).

Exercise 2. Cones and their arrows (on a fixed diagram) form a comma category. Show this and use this representation to define limits as functors from a category of diagrams to the category of cones.

Exercise 3. The category of graphs defined above has products (more generally, has all finite limits and colimits). Write a program to calculate products in this category.

5.2 Colimits in comma categories

The reader who attempted the last exercise, explicitly programming products of graphs, will have noticed that extensive use is made of products of sets. The set of nodes of the product graph is the product of the node-sets of the constituent graphs, likewise edges. Moreover, the product graph is assembled using the universal property of the product of sets.

This is part of a more general observation which gives a systematic method of calculating limits and colimits: Under certain circumstances, limits and colimits in comma categories are inherited from those in the constituent categories. This is the import of the following theorem, whose computational significance was pointed out to us by Goguen, see [Goguen, Burstall 84].

Theorem 3 *Let $L : \mathbf{A} \to \mathbf{C}$ and $R : \mathbf{B} \to \mathbf{C}$ be functors with L (finitely) cocontinuous. If $\mathbf{A}$ and $\mathbf{B}$ are (finitely) cocomplete so is the comma category (L, R).*

Proof Let Δ be a (finite) diagram in (L, R). Denote the object at node $n \in node(\Delta)$ by

$$\Delta_n = (a_n, f_n : L(a_n) \to R(b_n), b_n).$$

The projection functors *left* and *right* applied to Δ give diagrams Δ^A and Δ^B in categories $\mathbf{A}$ and $\mathbf{B}$. Let $\xi_n : a_n \to a$, $n \in node(\Delta)$ be the colimit of Δ^A in $\mathbf{A}$ and $\nu_n : b_n \to b$, $n \in node(\Delta)$ be the colimit of Δ^B in $\mathbf{B}$. Then $R(\nu_n)f_n : L(a_n) \to R(b)$, $n \in node(\Delta)$ is a cocone on $L(\Delta^A)$. The colimit of this diagram, by the cocontinuity of L, is the cocone $L(\xi_n) : L(a_n) \to L(a)$, $n \in node(\Delta)$. Thus there is a unique arrow $f : L(a) \to R(b)$ such that for all nodes $n \in nodes(\Delta)$ the following commutes:

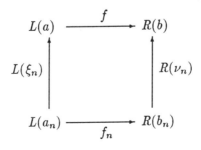

It may be verified that

$$(\xi_n, \nu_n) : \Delta_n \to (a, f, b), \quad n \in node(\Delta)$$

is a colimiting cocone on Δ with the universal property determined by
those of colimits ξ_n and ν_n. $\square$

Notice how the colimiting cocone is constructed from the universality
of colimits. To program this construction, we therefore need the universality as a component of the colimit structure. We have already made
this available, representing universality as a function.

The program derived from this construction takes as arguments two
functors L and R, colimits `colimA` in category A and `colimB` in B and
also the preservation property **preserve** of the the cocontinuous functor
L. The preservation property is a function taking colimiting cocones to
colimiting cocones. The result of the computation is a colimiting cocone
`colim_cocone` of a diagram D in the comma category (L,R), calculated
as follows:

```
let val (c_A,univ_A)= colimA(apply_Fun_Diag(left(L,R),D)
    val (c_B,univ_B)= colimB(apply_Fun_Diag(right(L,R),D)
        { colimits in A and B of projections of D }
    val (Lc,Luniv) = preserve(c_A,univ_A)
        { colimiting cocone in C using cocontinuity }
    val Rc = apply_Fun_CoCone(R,c_B)
    val c1 =
      cocone(co_apex(Rc),
             base(Lc),
             fn n =>
               let val (_,f_n,_) =
                 obj_at_node(D)(n) in
               compose(range(L))(sides(Rc)(n),f_n) end)
        { constructing new cocone in C from cocone Rc }
    val f = co_apex_arrow(Luniv c1)
        { using universality to construct arrow }
    val colim_obj = (co_apex(c_A),f,co_apex(c_B)) in
cocone(colim_obj,
       D,
       fn n =>
         comma_arrow( obj_at_node(D)(n),
                      (sides(c_A)(n),sides(c_B)(n)),
                      colim_obj) ) end
```

The universal part of this cocone is a function taking a cocone in the comma category and yielding a cocone arrow in the same category. It is computed using the universality of the colimits in the categories A and B.

```
fn c =>
  let val uA =
    co_apex_arrow(univ_A(apply_Fun_CoCone(left(L,R),c)))
      val uB =
    co_apex_arrow(univ_B(apply_Fun_CoCone(right(L,R),c)))
  in cocone_arrow( colim_cocone,
                   comma_arrow( colim_obj,
                                (uA,uB),
                                co_apex(c) ),
              c ) end
```

These two expressions combine to give a function lift_colimit which calculates colimits in comma categories:

```
fun cocomplete_comma_cat(cA,cB)
        (cL as cocontinuous_functor(L,_),R) =
  cocomplete_cat(comma_cat(L,R),lift_colimit(cA,cB)(cL,R))
```

5.3 Calculating colimits of graphs

Graphs, unlike sets, may not, in general, be decomposed as coproducts. What is required for graphs is a decomposition which takes account of shared subgraphs. Pushouts, rather than coproducts, provide a way of constructing and decomposing graphs. The shared part is the common source in a pushout diagram. More generally, decomposing objects with internal structure involves shared sub-structure. The sharing can often be described using arrows between sub-components. The decomposition is then that of a colimit on the diagram of sub-components. Pushouts have a special role, providing a semantics for parameter passing. This was first pointed out in the language Clear [Burstall, Goguen 80] and is discussed in Chapter 9.

We use the above program to compute a pushout of graphs. We have shown how the category of graphs **FinGraph** is (isomorphic to) the comma category $(I_{\textbf{FinSet}}, \mathcal{X})$ where $\mathcal{X} : a \mapsto a \times a$. The conditions of the theorem are satisfied since **FinSet** is finitely cocomplete and identity functors are cocontinuous. The finite cocompleteness of the category of graphs is therefore given by the following expression:

```
val cocomplete_cat_of_graphs =
    let val C = cocomplete_FinSet in
  cocomplete_comma_cat(C,C)( cocontinuous_I(FinSet),
                             cross_product) end
```

Let us run an example of a pushout in **FinGraph**. We compute the pushout of two graph arrows from the following graph, a_graph,

to the graphs below.

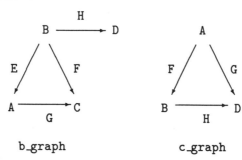

b_graph   c_graph

Let the arrow from a_graph to b_graph be the inclusion and that from a_graph to c_graph be defined as follows. The map on nodes is $\{A \mapsto B, B \mapsto A, C \mapsto B\}$ and that on edges is $\{E \mapsto F, F \mapsto F\}$.

We first define this diagram in **FinGraph** and then show the result of the computation.

The shape of a pushout diagram is a span:

To define the diagram we assign graphs to nodes and graph arrows to edges. The graph at node "a" is:

```
val a_graph =
    let val Na = {just "A",just "B",just "C"}
        val Ea = {just "E",just "F"} in
```

```
( Ea,
  set_arrow(Ea,fn just "E" => pair(just "B",just "A")
             | just "F" => pair(just "B",just "C"),
        cross_product ofo Na),
  Na ) end
```

The other graphs are defined similarly. The graph arrow associated with "f" is the inclusion, whilst that associated with "g" is expressed as the following arrow in a comma category:

```
val g_arrow =
      let fun gN_fn (just "A") = just "B"
          | gN_fn (just "B") = just "A"
          | gN_fn (just "C") = just "B"
        fun gE_fn (just "E") = just "F"
          | gE_fn (just "F") = just "F"
      in comma_arrow( a_graph,
                    ( set_arrow(edges(a_graph),
                                gE_fn,
                                edges(c_graph)),
                      set_arrow(nodes(a_graph),
                                gN_fn,
                                nodes(c_graph)) ),
                    c_graph ) end
```

The diagram is assembled as follows:

```
val D =
      let fun nodes_to_objs (word "a") = a_graph
          | nodes_to_objs (word "b") = b_graph
          | nodes_to_objs (word "c") = c_graph
        fun edges_to_arrows (word "f") = f_arrow
          | edges_to_arrows (word "g") = g_arrow
      in diagram(span,nodes_to_objs,edges_to_arrows) end
```

The computed value of the expression,

```
let val cocomplete_cat(_,graph_colimit) =
cocomplete_cat_of_graphs in graph_colimit(D) end
```

is a cocone and its universal property. The apex of the cocone is a graph whose nodes are the following set:

{pink just "B", pink just "A",
   blue pink just "D", blue blue pink just "D"}

The edges are as follows:

{pink just "E", blue pink just "H", blue pink just "G",
   blue blue pink just "H", blue blue pink just "G"}

The sources and targets of these edges are defined by the following function.

```
fn pink just "E" =>
     pair(pink just "B",pink just "A")
 | blue pink just "H" =>
     pair(pink just "B",blue pink just "D")
 | blue pink just "G" =>
     pair(pink just "A",pink just "A")
 | blue blue pink just "H" =>
     pair(pink just "A",blue blue pink just "D")
 | blue blue pink just "G" =>
     pair(pink just "B",blue blue pink just "D")
```

We draw this graph below (abbreviating the labels):

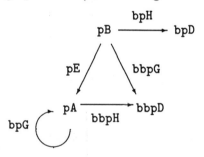

How does it get the answer in this form? The program for the inheritance of colimits by comma categories is invoked. This calls the calculation of colimits in **FinSet**. These colimits are calculated using the colimit extension function of the previous chapter and the code for initial object, binary coproducts and coequalizers in **FinSet**. The set of nodes and the set of edges of the pushout graph are both calculated as pushouts in **FinSet**. Thus the labelling with 'pink' and 'blue' of the disjoint union takes place as well as the selection of elements associated with the coequalizing. The graph is assembled by defining the source and target of edges using the universality of the pushout of the edges.

As well as the pushout graph, the arrows into it from the diagram, and the universality of the construction, are calculated. We leave it to the reader to work out what these are.

5.4 Functor categories

The class of functors between two categories can be considered to be a category by introducing arrows between functors called *natural transformations*. Natural transformations are functions which exhibit a form of parametric polymorphism in which the uniformity of the definition over different objects is expressed by commuting diagrams.

We define natural transformations and functor categories and give computational representations. We then consider calculating colimits in functor categories much as we did for comma categories.

5.4.1 Natural transformations

We begin with a definition.

Definition 19 *Let $F, G : \mathbf{A} \to \mathbf{B}$ be functors, a natural transformation $\alpha : F \to G$ is a function assigning to each object a of $\mathbf{A}$ an arrow of $\mathbf{B}$, $\alpha(a) : F(a) \to G(a)$, such that, for every arrow $f : a \to a'$ in $\mathbf{A}$ the following square commutes.*

$$
\begin{array}{ccc}
F(a) & \xrightarrow{\ \alpha(a)\ } & G(a) \\
\downarrow{\scriptstyle F(f)} & & \downarrow{\scriptstyle G(f)} \\
F(a') & \xrightarrow[\ \alpha(a')\ ]{} & G(a')
\end{array}
$$

Examples

Consider the two functors $\mathcal{X}, I : \mathbf{Set} \to \mathbf{Set}$ where $\mathcal{X}(a) = a \times a$ and similarly on arrows. There is a natural transformation $\delta : I \to \mathcal{X}$ defined for a set a as the function $\delta(a) = \lambda x.(x, x)$. That this is natural means that the following square commutes for any arrow $f : a \to a'$:

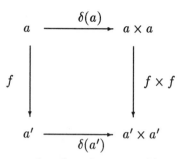

Polymorphic list processing functions provide examples of natural transformations. There is a functor $(-)^* : \textbf{Set} \to \textbf{Set}$ taking a set to the set of finite linear lists on the set, and operating as 'maplist' on arrows. Consider the reverse function on lists $rev(a) : a^* \to a^*$. It is a natural transformation as the following square commutes for any $f : a \to b$:

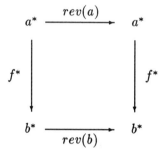

Let b^a be the set of functions from set a to set b. The evaluation function, $eval : b^a \to a$ defined simply as $eval(h, a) = h(a)$ is natural, i.e. defines a natural transformation, as it satisfies the equation $f(eval(h, x)) = eval(fh, x)$.

Adjunctions, introduced in the next chapter, provide a general setting for these examples.

5.4.2 Functor categories

We define a composition of natural transformations so that they form arrows in a category whose objects are functors. In fact, there are two compositions of natural transformations:

1. Let $F, G, H : \textbf{A} \to \textbf{B}$ be functors and $\alpha : F \to G$ and $\beta : G \to H$ be natural transformations. Define the *vertical composition, $\beta.\alpha$,* by, for each object a of $\textbf{A}$:

$$(\beta.\alpha)(a) = \beta(a)\alpha(a)$$

2. Let $F, F' : \mathbf{A} \to \mathbf{B}$ and $G, G' : \mathbf{B} \to \mathbf{C}$ be functors and $\alpha : F \to F'$ and $\beta : G \to G'$ be natural transformations. The *horizontal composition*, $\beta \circ \alpha$ (or simply $\beta\alpha$), is defined to be, for each object a of $\mathbf{A}$, the diagonal of the following commuting square.

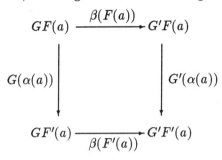

The notation used for these two compositions should be noted.

There are identity natural transformations on each functor F, ι_F, defined by for each object a of $\mathbf{A}$, $\iota_F(a) = i_{F(a)}$. Using identities, we define compositions of functors with natural transformations. For example $G\alpha = \iota_G\alpha$.

Definition 20 *The* functor category *from category* $\mathbf{A}$ *to category* $\mathbf{B}$, $\mathbf{B}^{\mathbf{A}}$, *has functors from* $\mathbf{A}$ *to* $\mathbf{B}$ *as objects and natural transformations as arrows. Composition is the vertical composition of natural transformations and identities are defined above.*

Examples

Categories of relations treated as spans form functor categories. Each span in a category $\mathbf{C}$ is a functor from the finite category whose shape is that of a span to the category $\mathbf{C}$. Considering graphs to be parallel pairs of arrows, the category of graphs **FinGraph** is a functor category. Thus graphs can be viewed not only as objects in a comma category but also as functors. This gives an alternative treatment to that of the previous section. Other examples of functor categories are categories of diagrams, categories of algebras [Lawvere 63], the category of 'sets through time' and other time structures of temporal logic, as well as representations of algebras (e.g. groups as permutations). Categories of functors find application in formalizing aspects of programming, e.g. the semantics of block structure [Oles 85].

Representation

We now represent natural transformations as a type and the functor category construction as a function. Natural transformations are triples consisting of the source and target functors and a map from objects to arrows:

```
datatype ('oA,'aA,'oB,'aB)Nat_transform =
    nat_transform of ('oA,'aA,'oB,'aB)Functor *
                     ('oA->'aB) *
                     ('oA,'aA,'oB,'aB)Functor
```

The category of functors is a function taking a pair of categories and returning a category:

```
fun id(A,cat(_,_,i,_)) F =
      nat_transform(F,fn a => i(F ofo a) ,F)
fun dotcomp(A,cat(_,_,_,comp)) =
      fn (nat_transform(_,beta,H),
          nat_transform(F,alpha,_)) =>
      nat_transform(F,fn a => comp(beta(a),alpha(a)),H)
fun ringcomp(cat(_,_,_,comp)) =
      fn (nat_transform(G,beta,G'),
          nat_transform(F,alpha,F')) =>
    nat_transform(G Fun_comp F,
                  fn a =>
                    comp(beta(F' ofo a),G ofa alpha(a)),
                  G' Fun_comp F')
fun cat_of_functors(A,B) =
      cat( fn nat_transform(s,_,_) => s,
           fn nat_transform(_,_,t) => t,
           id(A,B),
           dotcomp(A,B) )
```

5.5   Colimits in functor categories

Colimits in functor categories are derived from those in the target category. This is the import of the following theorem.

Theorem 4 *If the category* **B** *is (finitely) cocomplete, then, for any category* **A**, *the functor category* **B**^{A} *is (finitely) cocomplete.*

Proof The construction of colimits is pointwise. Let **B** be cocomplete and Δ be a diagram in $\mathbf{B}^{\mathbf{A}}$ with object Δ_n at $n \in node(\Delta)$. Apply diagram Δ to an object a of **A** to yield a diagram $\Delta(a)$ in **B**. Let the colimit of $\Delta(a)$ be

$$\xi_n(a) : \Delta_n(a) \to F(a), \quad n \in node(\Delta).$$

We show that F is a functor and that for each $n \in node(\Delta)$ the map $\xi_n \colon \Delta_n \to F$ is a natural transformation. Let $f : a \to a'$ in **A**. Using the universality of the colimit define $F(f) : F(a) \to F(a')$ as the unique arrow such that for all $n \in node(\Delta)$ the following square commutes:

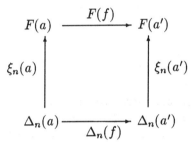

The cocone $\xi_n : \Delta_n \to F$ for $n \in node(\Delta)$ is colimiting, for let $\nu_n : \Delta_n \to G$, $n \in node(\Delta)$ be a cocone on Δ then for any object a of **A**, $\nu_n(a) : \Delta_n(a) \to G(a)$, $n \in node(\Delta)$ is a cocone on $\Delta(a)$ and thus there is a unique $\nu(a) : F(a) \to G(a)$ such that, for all $n \in node(\Delta)$, the following triangle commutes:

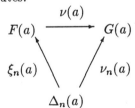

It can be verified that $\nu : F \to G$ is the unique natural transformation required. $\quad\square$

The derivation of a program from this proof is straightforward. We exhibit a fragment of the program – the construction of the colimit object of a diagram of functors. This object itself is a functor. The rest of the construction of the colimit we leave as an exercise.

```
fun colimiting_functor( A as cat(s,t,_,_),
                   cB as cocomplete_cat(B,_))(D) =
         { take an object of A, apply the
```

```
              diagram of functors and take colimit in B }
    let val object_function =
        fn a => colimit_object(cB)(applydo(A,B)(D,a))
        { the arrow part }
      val arrow_function =
        fn f =>
          let val univ =
            universal_part(cB)(applydo(A,B)(D,s(f)))
              val tc =
            colimit_cocone(cB)(applydo(A,B)(D,t(f)))
              val c1 =
            new_cocone(B)(applyda(A,B)(D,f),tc)
          in co_apex_arrow(univ c1) end
    in ffunctor(A,object_function,arrow_function,B) end
```

Here we use functions `applydo` and `applyda`, applying a diagram of functors to an object and to an arrow respectively, and the function `new_cocone` taking a cocone and a diagram arrow into its base, and producing a new cocone by composition.

5.6   Duality and limits

Limits in **C** are colimits in *dual*(**C**). Thus to compute limits in comma categories, we compute colimits in the dual of comma categories. How do we do this? The answer lies in the existence of isomorphisms which express the interaction of duality with category constructions, as in the following proposition:

Proposition 5 *There is an isomorphism*

$$\psi : dual(L, R) \to (dual(R), dual(L))$$

satisfying the equations $dual(left) = right.\psi$ *and* $dual(right) = left.\psi$.

We define this isomorphism as two, mutually-inverse, functors:

```
fun I1(L,R) =
   let val object_fn = fn (c,f,a) => (a,f,c) in
ffunctor(comma_cat(dual_Fun(R),dual_Fun(L)),
         object_fn,
         fn comma_arrow(S,(mc,ma),T) =>
           comma_arrow(object_fn(T),(ma,mc),object_fn(S)),
         dual_Cat(comma_cat(L,R)) ) end
```

```
fun I2(L,R) =
  let val object_fn = fn (a,f,c) => (c,f,a) in
ffunctor(dual_Cat(comma_cat(L,R)),
        object_fn,
        fn (comma_arrow(S,(ma,mc),T)) =>
         comma_arrow(object_fn(T),(mc,ma),object_fn(S)),
        comma_cat(dual_Fun(R),dual_Fun(L)) ) end
```

To compute colimits in the dual of a comma category, we compute colimits in a comma category and transport them along an isomorphism. Thus limits in comma categories can be computed from colimits in comma categories. To program this we use a simple function iso_colimit which takes an isomorphism of categories and yields a map of colimiting cocones. The following expression computes limits in the comma category (L,R) where R is finitely continuous.

```
        { first compute colimit in (dual(R),dual(L)) }
let val colim =
  lift_colimit(dual_complete_Cat(cB),dual_complete_Cat(cA))
              (dual_continuous_Fun(cR),dual_Fun(L)) in
      { apply iso to get colimit in dual(L,R) }
    val iso_colim = iso_colimit(I1(L,R),I2(L,R))(colim)
        { then convert to a limit in (L,R) }
in dual_colimit(iso_colim) end
```

As an example, we give a program which computes limits of graphs. Firstly, we need the finite continuity of the product functor $\mathcal{X}$ taking a finite set a to $a \times a$.

Exercise 4. Write a program expressing the product functor $\mathcal{X} : a \mapsto a \times a$ as a finitely continuous functor cts_cross_product i.e. show how $\mathcal{X}$ preserves finite limits. This may be coded from first principles or, alternatively, using a general result which says that functors defined in terms of limits are continuous.

To calculate limits of graphs use the following: (1) the finitely continuous functor cts_cross_product, (2) the finite completeness of **FinSet**, which we encoded in the previous chapter; and (3) the above limit inheritance which is a function complete_comma_cat taking a pair of complete categories and two functors, one of which is finitely continuous, and yielding a complete comma category. Combining these, we get the finite completeness of **FinGraph**:

```
val complete_cat_of_graphs =
    let val cC = complete_FinSet in
complete_comma_cat(cC,cC)(I(FinSet),cts_cross_product) end
```

For limits in functor categories, use the following proposition and the same method as for comma categories.

Proposition 6 *There is an isomorphism*

$$\theta : dual(\mathbf{B^A}) \rightarrow dual(\mathbf{B})^{dual(\mathbf{A})}$$

Exercise 5. The isomorphisms linking duality and constructions of categories are canonical. There is a unique functor (which is an isomorphism) satisfying the conditions of the propositions and an additional condition arising from the universal characterization of comma categories and functor categories. Fill out the details and prove this.

Exercise 6. Write a program to compute limits in functor categories.

5.7   Abstract colimits and limits*

The reader may have noticed that, in the previous chapter, initial objects, binary coproducts, coequalizers and finite colimits are all represented by types of the same general form. The relevant observation from category theory is that these concepts are all cases of a more general concept – that of colimits – and by choosing the appropriate diagrams the special colimits above can be realized.

Colimits are defined in terms of representations – representations of diagrams and of cocones. Since much of the manipulation associated with colimits is independent of the representation of diagrams and cocones, there is good reason for dispensing with the representation and providing an abstract description: an abstract type for colimits. The faint-hearted may complain that colimits are already abstract enough for their tastes! However, this is more than a mathematical exercise: it will produce code of added generality – provided it can be accommodated within the type structure. For instance, the construction of colimits in comma categories works equally well for initial objects and binary coproducts. Thus we need to pass as argument to the code the relevant colimit structure – an instance of the abstract type describing the colimits with which we are concerned.

The abstraction which we define contains the functorial nature of the construction of categories of diagrams and categories of cocones together with the function which, given a cocone, yields its base. Colimits are described as free objects. We call the structures *abstract colimit structures*.

In the following we consider the category **Cat** of categories as a 2-category (defined in Chapter 3, Exercise 7). This means that functors $\Sigma : \mathbf{Cat} \to \mathbf{Cat}$ are 2-functors acting not only upon the objects (categories) and arrows (functors) but also upon the natural transformations in such a way as to preserve identity and both compositions of natural transformations. Moreover, a natural transformation between functors on **Cat**, $\alpha : \Sigma \to \Sigma'$, respects natural transformations in **Cat**, $\gamma : F \to G$, where $F, G : \mathbf{A} \to \mathbf{B}$, in that the following holds:

$$\Sigma'(\gamma)\alpha(\mathbf{A}) = \alpha(\mathbf{A})\Sigma(\gamma)$$

The abstraction is phrased in terms of adjunctions. Those not familiar with adjunctions should turn to the next chapter, before reading on.

5.7.1 Abstract diagrams and colimits

An *abstract cocone structure* consists of a pair of functors $\Gamma, \Delta : \mathbf{Cat} \to \mathbf{Cat}$ with a natural transformation $\beta : \Gamma \to \Delta$. For each category **C**, call $\Delta(\mathbf{C})$ the category of *diagrams* (of fixed shape) on **C**, $\Gamma(\mathbf{C})$ the category of *cocones* on **C** and $\beta(\mathbf{C})$ the *base* functor.

If $\beta : \Gamma \to \Delta$ is a cocone structure, a *colimit* in category **C** of an object d of category $\Delta(\mathbf{C})$ is a free object $(\gamma, f : d \to \beta(\mathbf{C})(\gamma))$ on d with respect to the functor $\beta(\mathbf{C})$. A category **C** with such free objects for all objects of $\Delta(\mathbf{C})$, i.e. such that there is a left adjoint $H : \Delta(\mathbf{C}) \to \Gamma(\mathbf{C})$ to the functor $\beta(\mathbf{C}) : \Gamma(\mathbf{C}) \to \Delta(\mathbf{C})$ is called β-*cocomplete*.

A functor $F : \mathbf{A} \to \mathbf{B}$ is said to preserve the colimit of an object d of $\Delta(\mathbf{A})$, (γ, f), if $((\Gamma(F))(\gamma), (\Delta(F))(f))$ is a colimit of $(\Delta(F))(d)$. A functor preserving all existing colimits is called β-*cocontinuous*. For β-cocomplete categories **A** and **B**, $F : \mathbf{A} \to \mathbf{B}$ being β-cocontinuous means that the following is an arrow of adjunctions (see Chapter 6, Exercise 11, for definition):

$$\langle \Gamma(F), \Delta(F) \rangle : (H, \beta(\mathbf{A}), \eta) \to (H', \beta(\mathbf{B}), \eta')$$

An *abstract colimit structure* (β, H) on a category **C** is an abstract cocone structure $\beta : \Gamma \to \Delta$ with H a left adjoint to $\beta(\mathbf{C})$. Changing left adjoint to right adjoint, the above treatment defines an abstract *limit* structure.

5.7.2   Category constructions

Considering diagrams of a fixed shape $\Delta : \mathbf{Cat} \to \mathbf{Cat}$ and cocones on a fixed shaped base $\Gamma : \mathbf{Cat} \to \mathbf{Cat}$ we notice that for functors $L : \mathbf{C} \to \mathbf{A}$ and $R : \mathbf{C} \to \mathbf{B}$ there are isomorphisms,

$$\Delta(L, R) \cong (\Delta(L), \Delta(R))$$

$$\Gamma(L, R) \cong (\Gamma(L), \Gamma(R))$$

Moreover, a universal characterization of comma categories ensures that these isomorphisms uniquely satisfy certain commutation conditions. For the other constructions of categories that we have considered, there are similar isomorphisms:

$$\Delta(\mathbf{A} \times \mathbf{B}) \cong \Delta(\mathbf{A}) \times \Delta(\mathbf{B})$$

$$\Delta(\mathbf{B^A}) \cong \Delta(\mathbf{B})^{\mathbf{A}}$$

(and likewise for Γ) which, again using universal characterization of these constructions, uniquely satisfy certain commutation conditions. For these isomorphisms to exist, it is essential that we fix the shape of the diagrams under consideration.

We now insist that cocone structures satisfy these additional 'preservation' conditions so that we can formulate the construction of colimits in comma, functor and product categories.

It may be of interest to see these isomorphisms in a general setting. Consider a pair of functors $\Sigma, \Sigma' : \mathbf{Cat} \to \mathbf{Cat}$, then

- Define a *construction* on a category $\mathbf{A}$ to be a functor $H : \Sigma(\mathbf{A}) \to \Sigma'(\mathbf{A})$.

- Say an endofunctor $F : \mathbf{A} \to \mathbf{A}$ *respects* such a construction on $\mathbf{A}$ iff the following commutes *to within an isomorphism*.

$$
\begin{array}{ccc}
\Sigma(\mathbf{A}) & \xrightarrow{\ \Sigma(F)\ } & \Sigma(\mathbf{A}) \\
\Big\downarrow{\scriptstyle H} & & \Big\downarrow{\scriptstyle H} \\
\Sigma'(\mathbf{A}) & \xrightarrow[\ \Sigma'(F)\ ]{} & \Sigma'(\mathbf{A})
\end{array}
$$

As an example, for the product of categories, take Σ to be the functor such that $\Sigma(\mathbf{C}) = \mathbf{C}^{\underline{T}}$, $\underline{T}$ being the category of two objects and only identity arrows. The functor Σ' is the construction of the spanning category. i.e. for each category $\mathbf{C}$, $\Sigma'(\mathbf{C}) = \mathbf{C}^{\underline{Z}}$ where $\underline{Z}$ is the finite category below.

Let $\mathbf{A} = \mathbf{Cat}$ (thus interpreting the above in a higher universe). The functor H takes a pair of categories $\mathbf{A}$ and $\mathbf{B}$ to the product span $(first : \mathbf{A} \times \mathbf{B} \to \mathbf{A}, second : \mathbf{A} \times \mathbf{B} \to \mathbf{B})$. The existence of the isomorphisms above can be restated as: both the diagram functor Δ and the cocone functor Γ respect the product category construction.

Comma and functor categories fit this framework in a similar manner.

5.7.3 Indexed colimit structures

The considerations above restrict us to diagrams of a fixed shape and cocones on bases of a fixed shape. This treatment may, however, be extended to cover collections of colimits on variously shaped diagrams, such as finite coproducts, finite colimits or filtered colimits. Consider a class G (of graphs) and a G-indexed collection of cocone structures $\beta_g : \Gamma_g \to \Delta_g$, $g \in G$ such that for each $g \in G$ the abstract cocone structure respects the category constructions above. A category $\mathbf{C}$ is said to be G-cocomplete (with the cocone structures implicit) if, for each $g \in G$, the functor $\beta_g(\mathbf{C}) : \Gamma_g(\mathbf{C}) \to \Delta_g(\mathbf{C})$ has a left adjoint.

5.7.4 Discussion

The intention of the exercise was to capture enough of the behaviour of diagrams, cocones and colimits to enable us to formulate the construction of colimits (and limits) in product categories, comma categories and functor categories as well as proofs such as the cocontinuity of left adjoints and interchange of colimits. It may be verified that these proofs can be expressed within the simple colimit structures defined above.

Looking back over the code for colimits in comma and functor categories, we see that the abstract colimit structures contain just the auxiliary functions that occurred and therefore can be used to organize the programs as well as to produce code of greater generality.

There are several aspects of colimits which have not been included as they play no role in these proofs although they may well be needed elsewhere and may then be added to the abstract structure. For instance, the base of a colimit on a diagram d is itself d (i.e. $\beta(\mathbf{C})H$ is an identity). Moreover, arrows between diagrams of different shapes, for which there are several possible definitions, are not needed for the present purposes.

5.8   Exercises

Exercise 7. Product categories The product of categories $\mathbf{A}$ and $\mathbf{B}$, $\mathbf{A} \times \mathbf{B}$, has as objects pairs of objects and as arrows, pairs of arrows. Program up this construction and show how to compute limits and colimits componentwise.

Exercise 8. Twisted arrow categories For any category $\mathbf{C}$, we define a category whose objects are arrows of $\mathbf{C}$. Arrows in the category from $f : a \to b$ to $f' : a' \to b'$ are pairs of arrows $h : a \to a'$ and $k : b' \to b$ such that the following square commutes:

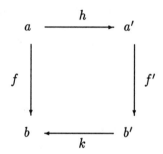

Composition of arrows is $(h, k).(h', k') = (hh', k'k)$. Express this construction as an ML program.

Exercise 9. Consider colimits in the comma category $(K_a, I_\mathbf{A})$ where K_a is the constant functor returning object a of $\mathbf{A}$. Objects in this category are arrows out of a. This is a case not considered so far (why?). Show that *coproducts* in this category can be constructed as *pushouts* in $\mathbf{A}$. Write the associated program.

Exercise* 10. Automata Non-deterministic finite automata are edge-labelled graphs. These are objects in the comma category,

$$(Edges : \mathbf{FinGraph} \to \mathbf{FinSet}, K_a : \underline{1} \to \mathbf{FinSet})$$

where K_a is the constant functor from the one object category returning the set a of labels. Show this and compare colimits in this category with constructions of non-deterministic finite automata for recognizing regular expressions.

Deterministic finite automata are based on a transition function of the form $Q \times A \to Q$ where Q is the set of states and A the input alphabet. These are examples of algebras, which we consider in the next chapter. Colimits of deterministic automata require a more complex construction than that of non-deterministic automata. A suitable reference for this material is [Hopcroft, Ullman 79].

Exercise* 11. Graph grammars Graph grammars use pushouts in assembling and matching graphs. If you are unfamiliar with this material consult references (e.g. [Ehrig et al. 73]). Write programs for the operations on graphs used in graph grammars.

Exercise* 12. Indexed categories An indexed category is a functor $F : \mathbf{C} \to \mathbf{Cat}$. From this we may create a category $\mathcal{G}(F)$, the opfibration induced by F, by what is known as the Grothendieck construction:

- Objects of $\mathcal{G}(F)$ are pairs $\langle c, a \rangle$ where c is an object of $\mathbf{C}$ and a is an object of $F(c)$.

- Arrows of $\mathcal{G}(F)$ are pairs $\langle f, g \rangle : \langle c, a \rangle \to \langle c', a' \rangle$ where $f : c \to c'$ in $\mathbf{C}$ and $g : (F(f))(a) \to a'$ in $F(c')$.

- The composition of $\langle f, g \rangle : \langle c, a \rangle \to \langle c', a' \rangle$ with $\langle f', g' \rangle : \langle c', a' \rangle \to \langle c'', a'' \rangle$ is given by

$$\langle f'f : c \to c'', g.'(F(f'))(g) : (F(f'f))(a) \to a'' \rangle.$$

Many categories are of the form $\mathcal{G}(F)$. For instance, let $\mathbf{C}$ be (the dual of) a category of theories and F allocate to each theory its variety of algebras. Other examples: signatures indexed by their sorts, theories (signatures and sets of equations) indexed by their signatures, diagrams by their shape (underlying graph), cones by their base (a diagram), etc. Notice how 'dependent' products arise as objects and arrows of $\mathcal{G}(F)$.

What about colimits in $\mathcal{G}(F)$? Establish the following:

Proposition 7 *If $\mathbf{C}$ has (finite) colimits and, for each object c of $\mathbf{C}$, $F(c)$ has (finite) colimits, and also, for each $f : c \to c'$ in $\mathbf{C}$, $F(f)$ preserves (finite) colimits, then $\mathcal{G}(F)$ has (finite) colimits.*

Limits are not quite as straightforward. Establish:

Proposition 8 *Suppose each functor $F(f)$ has a right adjoint. If* **C** *has (finite) limits and so do $F(c)$ for each object c of* **C**, *then* $\mathcal{G}(F)$ *has (finite) limits.*

This exercise was suggested by a draft paper of A. Tarlecki describing limit and colimit constructions in categories of the form $\mathcal{G}(F)$.

Chapter 6

Adjunctions

Adjunctions, introduced by Kan in 1958, provide a descriptive framework of great generality, capturing the essence of many canonical constructions. They turn up throughout mathematics often as 'closures' and 'completions', and as 'free' and 'generated' structures. For example, the transitive closure of a graph, the completion of a metric space, factor commutator groups, and free algebras are all examples of adjunctions. In categorical logic, quantifiers are interpreted as adjunctions with respect to substitution of variables. The canonical nature of these constructions is captured by universality. Indeed, adjunctions subsume the universal concepts of previous chapters, both limits and colimits.

Adjunctions arise in the semantics of programming languages. It was the 'ADJ' group (J.A. Goguen, J.W. Thatcher, E.G. Wagner and J.B. Wright) who showed that the the notion of a 'data type' involves initial and free algebras. Relevant papers are [Goguen, Thatcher, Wagner, Wright 77] and [Goguen, Thatcher, Wagner 78]. The construction of free algebras as term algebras links syntax with semantics; values of the data type are constructed as terms. Universality allows functions to be defined over the type. The dual notion, of cofreeness and terminal algebras, arises in the semantics of systems with internal states and an externally observed behaviour. For example, the minimal realization of the behaviour of finite automata determines an adjunction, see [Goguen 73], [Arbib, Manes 74].

In this chapter we define adjunctions and represent them as an ML type. Adjunctions are fairly complex structures and can be defined in several equivalent ways. We encode this equivalence as conversion functions between representations. We look at some examples of adjunctions, coding them as ML values. Moreover, we show how to compute with adjunctions, constructing continuous and cocontinuous functors.

At the end of the chapter, we present a categorical version of the 'term' or 'word' algebra construction of free algebras. This is an iterative construction which we use to construct the transitive closure of a graph. Another construction of adjunctions is provided by the adjoint functor theorems of Freyd [1964]. These are used to establish the existence of adjunctions. As programs for constructing adjunctions, they are of little use because the 'solution sets' from which adjunctions are constructed are necessarily finite.

6.1   Definitions of adjunctions

Adjunctions capture canonical constructions whereby objects of one kind are constructed from those of another kind in a universal manner. There are two categories involved in an adjunction, together with a pair of functors, $G : \mathbf{A} \to \mathbf{B}$ and $F : \mathbf{B} \to \mathbf{A}$. These functors are not inverse to one another but satisfy a universal property linking their composite with the identity.

We first introduce the concept of a universal arrow and then define adjunctions.

Definition 21 *Let* $G : \mathbf{A} \to \mathbf{B}$ *be a functor and* b *be an object of* $\mathbf{B}$, *a universal arrow from* b *to* G *is a pair* (a, u) *with* a *an object of* $\mathbf{A}$ *and* $u : b \to G(a)$, *such that, for any pair* $(a', f : b \to G(a'))$, *there is a unique arrow* $f^{\#} : a \to a'$ *such that the following triangle commutes:*

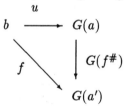

Definition 22 *Let* $\mathbf{A}$ *and* $\mathbf{B}$ *be categories. An* adjunction *is defined by either of the following equivalent definitions.*

1. *A functor* $G : \mathbf{A} \to \mathbf{B}$ *and, for each object* b *of* $\mathbf{B}$, *a universal arrow from* b *to* G.

2. *Two functors* $F : \mathbf{B} \to \mathbf{A}$ *and* $G : \mathbf{A} \to \mathbf{B}$ *together with two natural transformations* $\eta : I_{\mathbf{B}} \to GF$ *and* $\epsilon : FG \to I_{\mathbf{A}}$ *such that* $G\epsilon.\eta G = i_G$ *and* $\epsilon F.F\eta = i_F$.

The functor F is called the *left adjoint* to G, and G the *right adjoint* to F. The natural transformation $\eta : I_{\mathbf{B}} \to GF$ is the *unit* of the adjunction and $\epsilon : FG \to I_{\mathbf{A}}$ the *co-unit*.

We show how the second definition arises from the first. Let b be an object of $\mathbf{B}$ and (a, u) the corresponding universal arrow from b to G. Define the action of functor F on objects by $F : b \mapsto a$. Let $f : b \to b'$ in $\mathbf{B}$ and (a', u') be the universal arrow from b' to G. Define $F(f) = (u'f)^{\#} : a \to a'$ using the universality of (a, u).

The unit and co-unit are defined as follows. For b in $\mathbf{B}$ with universal arrow (a, u) define the unit $\eta(b) = u : b \to G(F(b))$. For the co-unit, let $(FG(a), u')$ be the universal arrow from $G(a)$ to G. Consider the pair $(G(a), i_{G(a)} : G(a) \to G(a))$, and define $\epsilon(a) = i_{G(a)}{}^{\#} : F(G(a)) \to a$.

The first equation arises from the commutativity of the triangle:

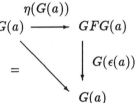

The second equation arises from the uniqueness property of the universal arrow. It is a straightforward exercise to verify that F is a functor and η and ϵ are natural transformations.

In the reverse direction, from the second definition to the first, define the universal arrow from b to G to be $(F(b), \eta(b))$. This is universal because, for any arrow $f : b \to G(a')$, define $f^{\#} = \epsilon(a')F(f) : F(b) \to a'$. We denote an adjunction by $\langle F, G, \eta, \epsilon \rangle : \mathbf{B} \to \mathbf{A}$. Notice that an adjunction defines a bijection between arrows:

$$\frac{b \to G(a)}{F(b) \to a}$$

The universality of adjunctions ensures that the right adjoint determines the left adjoint to within an isomorphism and *vice versa*.

The conversion between a universal definition and an equational definition is used to show the partial equational nature of many of the concepts of category theory. It also gives an evaluation mechanism for categorical expressions, treating the equations as rewrite rules. This arises in the categorical abstract machine [Curien 86] and in the categorical programming language of Hagino [1987], described in Chapter 10.

6.2   Representing adjunctions

Adjunctions are built from categories, functors, natural transformations
and universality. We have shown how to represent all of these as types in
ML. All that remains is to group them together to represent adjunctions.
The definition in terms of universal arrows gives the following type:

```
type ('oA,'aA,'oB,'aB)Universal_Arrow =
             'oB -> (('oA*'aB)*(('oA*'aB) -> 'aA))

datatype ('oA,'aA,'oB,'aB)Universal_Adj =
    universal_adj of ('oA,'aA,'oB,'aB)Functor *
                     ('oA,'aA,'oB,'aB)Universal_Arrow
```

The definition in terms of two natural transformations gives the following type:

```
datatype ('oA,'aA,'oB,'aB)Adjunction  =
    adjunction of ('oB,'aB,'oA,'aA)Functor *
                  ('oA,'aA,'oB,'aB)Functor *
                  ('oB,'aB,'oB,'aB)Nat_transform *
                  ('oA,'aA,'oA,'aA)Nat_transform
```

We write a program to convert from the first definition, in terms
of universal arrows, to the second, in terms of natural transformations.
The equivalence is explained in the previous section; the translation into
a program is straightforward. We first construct the left adjoint, and
then the two natural transformations:

```
fun left_adjoint(universal_adj(G,universal)) =
let val A = domain(G)
    val B = range(G)
    val obj_part =
     fn b => let val ((a,_),_) = universal(b) in a end
    val arrow_part =
     fn f =>
      let val (_,source_univ) = universal(source(B)(f))
          val ((a,u),_)       = universal(target(B)(f))
      in source_univ(a,compose(B)(u,f)) end
  in ffunctor(B,obj_part,arrow_part,A) end
```

```
fun UA_to_ADJ(u_adj as universal_adj(G,universal)) =
  let val A = domain(G)
      val B = range(G)
      val F = left_adjoint(u_adj)
      val unit =
        nat_transform
          ( I(B),
            fn b =>
            let val ((_,f),_) = universal(b) in f end,
            G Fun_comp F )
      val counit =
        nat_transform
          ( F Fun_comp G,
            fn a =>
            let val (_,univ) = universal(G ofo a)
            in univ(a,identity(B)(G ofo a)) end,
            I(A) )
  in adjunction(F,G,unit,counit) end
```

Exercise 1. The following is another definition of adjunctions, in terms of a bijection between arrows. An adjunction is a pair of functors $F : \mathbf{B} \to \mathbf{A}$ and $G : \mathbf{A} \to \mathbf{B}$ together with an isomorphism of comma categories

$$J : (F, I_{\mathbf{A}}) \to (I_{\mathbf{B}}, G)$$

such that $\langle left, right \rangle J = \langle left, right \rangle$. Prove that this is equivalent to the previous definitions. By defining an appropriate ML type, program this equivalence.

6.3 Examples

Adjunctions arise in many areas of mathematics. We begin with two simple examples as tutorials explaining in detail how the components of an adjunction arise. We then present the construction of free algebras as 'term' or 'word' algebras. Finally, to illustrate the wealth of constructions which are adjunctions, we list some other examples.

6.3.1   Floor and ceiling functions: converting real numbers to integers

Consider the two operations mapping real numbers to integers: $r\downarrow$ (called 'floor'), the largest integer less than r (e.g. $3.14\downarrow = 3$), and $r\uparrow$ (called 'ceiling'), the smallest integer greater than r (e.g. $3.14\uparrow = 4$). These simple operations determine adjunctions.

Consider the case of the ceiling function $r\uparrow$. The two categories involved are **int** and **real**, that of integers and that of real numbers, both as partial orders under the usual $\leq$ partial order. Recall that a partial order can be considered to be a category whose objects are elements of the partially ordered set, and for which there is an arrow from x to y just when $x \leq y$.

There is an inclusion functor, $U : \textbf{int} \to \textbf{real}$. The ceiling function is functorial$\uparrow$: **real** $\to$ **int**, because $r \leq r' \Rightarrow r\uparrow \leq r'\uparrow$. This functor is left adjoint to U. First notice that $r \leq U(r\uparrow)$; this provides the unit of the adjunction. The universality arises as follows. For any integer n, with $r \leq n$, i.e. $r \leq U(n)$ in **real**, we have that $r\uparrow \leq n$. This captures the defining characteristic of the ceiling function as the smallest integer greater than r. The triangle required to commute does so automatically in partial orders as there is at most one arrow between any two objects.

This example was brought to our attention by V. Pratt. Adjunctions between partial orders are exactly Galois connections, see [Mac Lane 71].

Exercise 2. Show that the floor function $r\downarrow$ is right adjoint to U.

Exercise 3. Express the two adjunctions defined by 'floor' and 'ceiling' as ML values. The representation of **int** and **real** can be drawn from Exercise 3 in Chapter 3 where it is suggested that an arrow $m \to n$, for $m \leq n$, is represented by $(m, n - m)$, or you may use the pair (m, n).

6.3.2   Components of a graph

Many adjunctions arise in graph theory. We select a simple example as an illustration, that of the components of a graph.

A *semipath* in a graph is a sequence of nodes, n_i, $1 \leq i \leq N$, such that for each pair of consecutive nodes n_i, n_{i+1}, there is an edge from n_i to n_{i+1} or from n_{i+1} to n_i. Two nodes in a directed multigraph are *connected* if there is a semipath linking them. A graph is said to be *connected* if all pairs of nodes are connected. A *connected component* of a graph is a maximal connected subgraph. Every graph can be expressed as a (disjoint) union of its connected components. The connection relation is the equivalence closure of the adjacency relation (two nodes being adjacent when they are linked by an edge).

We show how the connected components determine an adjunction. Denote a graph by (N, E, s, t), where N is the set of nodes, E the set of edges and s and t are the source and target maps. Let S be a set. We may form a graph from S, the discrete graph on S, as the graph $(S, S, \lambda x.x, \lambda x.x)$. This is functorial $D : \mathbf{Set} \to \mathbf{Graph}$. Any function $f : S \to S'$, gives a graph arrow $(f, f) : D(S) \to D(S')$.

We may label the nodes of a graph so that two nodes are in the same connected component iff they have the same label. Let S be a minimum set of labels, so that labels correspond exactly to components. This determines a functor $C : \mathbf{Graph} \to \mathbf{Set}$. The action on arrows is defined as follows. If $(f, g) : G \to G'$, f acting on nodes, g on edges and s is the component label of node n, then the image of s under $C(f, g)$ is the component label of $f(n)$. This is well defined as connected nodes map, under a graph arrow, to connected nodes.

We show that C is left adjoint to D. The unit of the adjunction is a graph arrow $\eta(G) : G \to DC(G)$ for each graph G. This is the function labelling each node with its component label. Edges are mapped to the unique edge on the node in the discrete graph.

The universal property of this construction arises as follows. Let S be a set and $(f_N, f_E) : G \to D(S)$ be a graph arrow. There is a function $f : C(G) \to S$ given by $f(s) = f_N(n)$ where n is in component s. Well-definedness follows from the fact that (f_N, f_E) is a graph arrow and that all elements of $C(G)$ are component labels.

Finally, f is unique making the following commute:

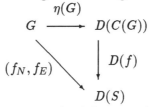

That f makes it commute is the definition of f, that it is unique follows from the fact that $\eta(G)$ does not map nodes in different components to the same label.

Informally, we may say that the existence part of the universality corresponds to $C(G)$ having 'no junk' (every element arises from a component of G), and the uniqueness part corresponds to $C(G)$ having 'no confusion' (every element arises from a unique component of G). This 'no junk, no confusion' is characteristic of many adjunctions.

To program this adjunction for finite sets and graphs, we need to compute the component functor C : **FinGraph** $\rightarrow$ **FinSet**. This requires a little thought. A neat way of computing this is to use our categorical programs. Recall that a graph can be interpreted as a parallel pair of arrows in **Set**. Their coequalizer corresponds to the set of components.

We first code the discrete graph functor:

```
val Discrete =
    let fun discrete_graph(S) =
               (S,S,identity_fn,identity_fn) in
  functor(FinSet,
          discrete_graph,
          fn set_arrow(S,f,S') =
           (discrete_graph(S),(f,f),(discrete_graph(S'))),
          FinGraph) end
```

Here is the adjunction determined by the components. The set of components is the coequalizer of the source and target arrows. The unit is the coequalizing arrow and the universality is that of the coequalizer.

```
val Components_Adj =
  let val univ_arrow =
    fn G as (N,E,s,t) =>
      let val ((S,q),univ) =
        coequalize(cocomplete_FinSet)
                   (set_arrow(E,s,N),set_arrow(E,t,N) in
  ((S,
    (G,(q,compose(q,set_arrow(E,s,N)))),Discrete ofo S)),
  fn (S',(_,(fe,fn),_)) =>
        univ(set_arrow(N,fn,S'))) end) in
        universal_adj(Discrete,univ_arrow)
```

None of this need be tied to the category of sets. In any category with coequalizers, we can define graphs as parallel pairs and discrete graphs as pairs of identity arrows and so construct the adjunction.

6.3.3 Free algebras

In Chapter 3, we considered some categories arising in universal algebra. We now show how the construction of term algebras (often called 'word' algebras) determines an adjunction. As well as its role in universal algebra, this is a fundamental idea in the semantics of data types and programming languages. We consider programs for this construction later in the chapter.

Recall that an operator domain is a set of operator symbols indexed by their arities (natural numbers). If Ω is an operator domain, denote by Ω_n the set of operators in Ω whose arity is the natural number n.

The terms in a set X over an operator domain Ω, the set of which we denote by $T_\Omega(X)$, are syntactic objects defined recursively by:

$$x \in X \Rightarrow \langle x \rangle \in T_\Omega(X)$$

$$\rho \in \Omega_n, \ t_1, t_2, \ldots, t_n \in T_\Omega(X) \Rightarrow \rho(t_1, t_2, \ldots, t_n) \in T_\Omega(X)$$

We can consider $T_\Omega(X)$ to be an Ω-algebra, by defining the operations syntactically. That is, for $\rho \in \Omega_n$, the corresponding operation takes the terms $t_1, t_2, \ldots, t_n \in T_\Omega(X)$ to the term $\rho(t_1, t_2, \ldots, t_n)$. This algebra is called the term algebra on X.

There is a functor $U : \mathbf{Alg}_\Omega \to \mathbf{Set}$ taking each algebra to its carrier (its underlying set) and acting as identity on arrows. The construction of the term algebra defines a left adjoint to U, namely $F : \mathbf{Set} \to \mathbf{Alg}_\Omega$, where $F(X)$ is $T_\Omega(X)$ considered as an algebra. We describe this by saying that the term algebra on X is a *free algebra* on X.

The universality arises as follows. There is a function $u : X \to U(T_\Omega(X))$, given by $u(x) = \langle x \rangle$. Now, for any Ω-algebra (A, α) and function $f : X \to A$, we define the homomorphism $f^\# : T_\Omega(X) \to (A, \alpha)$ recursively:

$$f^\#(\langle x \rangle) = f(x) \text{ for } x \in X$$

$$f^\#(\rho(t_1, t_2, \ldots, t_n)) = \rho(f^\#(t_1), f^\#(t_2), \ldots, f^\#(t_n))$$

It is a homomorphism by the second clause of the definition. The first clause ensures that the following triangle commutes:

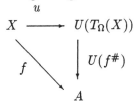

These two clauses therefore uniquely define such a homomorphism.

In the case of algebras of equational theories, we construct free algebras as quotients of term algebras as follows.

Let $\mathcal{E}$ be a set of Ω-equations. Define an equivalence relation $\sim$ on $T_\Omega(X)$ by $s \sim t$ iff $\mathcal{E} \vdash s = t$, that is, we can prove, using equational deduction from equations $\mathcal{E}$, that $s = t$ holds. Let $T_\Omega(X)/\mathcal{E}$ be the set of equivalence classes of $T_\Omega(X)$ under $\sim$. This is the carrier of an Ω-algebra where we define the operations as follows:

$$\rho([t_1], [t_2], \ldots, [t_n]) = [\rho(t_1, t_2, \ldots, t_n)]$$

Here $[t]$ denotes the equivalence class containing t. The operation is well defined by definition of the relation $\sim$.

A similar argument to that above shows that $T_\Omega(X)/\mathcal{E}$ is a free algebra, i.e. determines a left adjoint to $U : \mathbf{Alg_T} \to \mathbf{Set}$ where $\mathbf{T} = (\Omega, \mathcal{E})$.

We consider two examples of the above constructions, natural numbers and finite powersets.

The natural numbers $\mathcal{N} = \{0, 1, \ldots\}$ form an Ω-algebra where Ω contains just a constant c and a unary operator f. We interpret c as 0 and f as the successor function $\lambda x . x + 1$. The natural numbers are identified with terms in $T_\Omega(\phi)$ where ϕ is the empty set and so forms a free algebra. Free algebras on the empty set are initial objects in the category $\mathbf{Alg_\Omega}$.

As another example, let S be a set and $\mathcal{F}(S)$ denote the set of finite subsets of S. We show that this forms a free semilattice. A semilattice is an algebra with one binary operation that is associative, commutative, absorptive and has an identity. $\mathcal{F}(S)$ forms a semilattice under the operation $\cup$ of union of sets. Its identity is the empty set. The equivalence classes of terms in variables S are in 1-1 correspondence with finite subsets of S. An equivalence class of a term t contains all bracketed versions of t, all versions with the identity appearing, all permutations of the variables, and all multiplicities of variables. Therefore, all that is required to identify an equivalence class is the finite set of variables occurring in the term. An alternative proof that $\mathcal{F}(S)$ is a free semilattice can be given by demonstrating the universality of $\mathcal{F}(S)$.

6.3.4 Graph theory

Here are some examples of adjunctions in graph theory:

- The transitive closure of a graph is the graph with the same nodes but whose edges are paths in the original graph. A path in a graph is a (possibly empty) sequence of edges $e_1, e_2, \ldots, e_n$, $n > 0$ such that $t(e_i) = s(e_{i+1})$, $0 < i < n$, where s and t are the source and target, respectively, of the edges. A path into a node may be composed with one out of the node. Empty paths at each node are identities of this composition so that the transitive closure forms a category, the *path category* of the graph. The construction is functorial, T : **Graph** $\rightarrow$ **SmallCat** where **SmallCat** is the category whose objects are small categories (categories with a set of objects and a set of arrows) and whose arrows are functors. The functor T is left adjoint to the functor U : **SmallCat** $\rightarrow$ **Graph** which returns the underlying graph of a category.

- Two nodes, m and n, in a graph are said to be strongly-connected if there is a path from m to n and a path from n to m. A subgraph is strongly connected if each pair of nodes in it is strongly connected. A strong component of a graph is a maximal strongly connected subgraph. The strong components of a graph themselves form an acyclic graph which is a quotient of the original graph. This may be expressed as a left adjoint to the inclusion functor **AcyclicGraph** $\rightarrow$ **Graph** where **AcyclicGraph** is the full subcategory of **Graph** of acyclic graphs.

- Consider graphs whose edges have non-negative real numbers $\mathcal{R}^+$ assigned to them. The distance between two nodes is the shortest path (in terms of the sum of the distances on edges in the path) between the nodes. By allocating to each pair of nodes the distance between them (making it infinite if there is no path between them), we form a metric space from the graph. Two categories are involved: (1) **MetricSpace** whose objects are metric spaces and whose arrows $f : \langle S, d \rangle \rightarrow \langle S', d' \rangle$ satisfy $d(m, n) \geq d'(f(m), f(n))$; (2) **LabelledGraph** whose objects are graphs whose edges are labelled with non-negative real numbers and whose arrows are graph arrows $\langle f_N, f_E \rangle : (N, E, s, t, \lambda) \rightarrow (N', E', s', t', \lambda')$ where $\lambda : E \rightarrow \mathcal{R}^+$ and $\lambda' : E' \rightarrow \mathcal{R}^+$ satisfying $\lambda(e) \geq \lambda'(f_E(e))$.

There is a functor G : **MetricSpace** $\rightarrow$ **LabelledGraph** defined as follows. For any metric space $\langle S, d \rangle$, the graph $G(\langle S, d \rangle)$ has elements of S as nodes, and has an edge from m to n just when $d(m, n) < \infty$, the label being $d(m, n)$. The construction of a metric space from a graph by taking the minimum path distance between nodes provides a left adjoint to G.

6.3.5   Limits and colimits

Both limits and colimits can be interpreted as adjunctions.

For example, the initial object in a category C is a left adjoint to the constant functor $T : C \to 1$, where 1 is the category of one object and an identity arrow. The unit of the adjunction is the identity in 1 and the co-unit is the unique arrow from the initial object.

We can code this as a function converting initial objects to adjunctions:

```
datatype UnitObj   = unitobj
datatype UnitArrow = unitarrow
val unitcat = cat (fn unitarrow => unitobj,
                   fn unitarrow => unitobj,
                   fn unitobj => unitarrow,
                   fn (unitarrow,unitarrow)=> unitarrow)
val T(C) = ffunctor( C,
                   fn c => unitobj,
                   fn f => unitarrow,
                   unitcat)

fun initial_adj(C,(init,init_univ)) =
    universal_adj(T(C),
                  fn unitobj =>
                  ((init,identity(unitcat)(unitobj)),
                   fn (c,_) => init_univ(c)))
```

As another example, the reader may like to show that binary coproducts are left adjoint to the diagonal functor $\Delta : C \to C \times C$, $\Delta(a) = (a, a)$. The example of graph components (Example 5.3.2) shows how coequalizers form an adjunction. Limits arise as right adjoints. For example, products are right adjoint to Δ.

Exercise 4. General colimits in a category determine an adjunction between categories of diagrams and categories of cocones. Demonstrate this and write a function which constructs an adjunction from a (finitely) cocomplete category.

6.3.6   Adjunctions and comma categories

An important construction of an adjunction arises from slice categories, a special case of comma categories.

Recall that, for category **C** and object a in **C**, $\mathbf{C}/a$ denotes the comma category $(I_{\mathbf{C}}, K_a)$, where $K_a : \mathbf{1} \to \mathbf{C}$ is the constant functor returning the object a. Objects in $\mathbf{C}/a$ are arrows into a and arrows in $\mathbf{C}/a$ are commuting triangles:

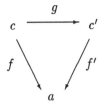

A function $h : a \to b$ defines a functor $\Sigma_h : \mathbf{C}/a \to \mathbf{C}/b$ by composition, taking $f : c \to a$ into $hf : c \to b$.

When **C** has pullbacks, this functor has a right adjoint $h^{\#} : \mathbf{C}/b \to \mathbf{C}/a$ defined as follows. For object $f : c \to b$, form the pullback:

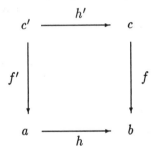

Then $h^{\#} : f \mapsto f'$. The universal property of pullbacks provides the action on arrows.

Exercise 5. Complete the description of this adjunction and express it as an ML program.

When $\mathbf{C} = \mathbf{Set}$, the category $\mathbf{C}/a$ is that of a-indexed sets. Describe the functors Σ_h and $h^{\#}$ in this case.

Exercise 6. When **C** has binary products, the functor $\Sigma_a : \mathbf{C}/a \to \mathbf{C}$, defined by $\Sigma_a : (f : c \to a) \mapsto c$, has a right adjoint $\times a : \mathbf{C} \to \mathbf{C}/a$ mapping c to the projection $\pi_2 : c \times a \to a$. Prove this and code up the construction.

6.3.7 Examples from algebra and topology

Textbooks on category theory give many examples of adjunctions drawn from algebra and topology. We list a few examples here.

In algebra, the factor commutator group is the free abelianization of a group, i.e. is left adjoint to the inclusion functor of abelian groups in **Group**. The field of quotients of an integral domain is left adjoint to the inclusion of fields in the category of integral domains (with monic maps). Many other examples arise as freely generated structures.

In topology, the functor taking a topological space to its underlying set has a left adjoint (the discrete topology on a set, all subsets are open sets) and a right adjoint (the indiscrete topology, only the set itself and the empty set are open sets). The Stone-Čech compactification of a topological space is left adjoint to the inclusion of compact spaces in the category of all topological spaces.

6.4   Computing with adjunctions

We now look at a construction involving adjunctions and show how to code it in ML. We start with a duality operation on adjunctions and then look at a construction of continuous and cocontinuous functors from adjunctions.

The *dual* of adjunction $\langle F, G, \eta, \epsilon \rangle : \mathbf{B} \to \mathbf{A}$ is the adjunction

$$\langle dual(G), dual(F), \epsilon, \eta \rangle : dual(\mathbf{A}) \to dual(\mathbf{B}).$$

```
fun dual_Adj(adjunction(F,G,unit,counit)) =
        adjunction(dual_Fun(G),dual_Fun(F),counit,unit)
```

A useful property of adjunctions is contained in the next proposition.

Proposition 9 *If $\langle F, G, \eta, \epsilon \rangle : \mathbf{A} \to \mathbf{B}$ is an adjunction then F is cocontinuous and G is continuous.*

Proof   We show that F is cocontinuous. Duality provides the other half of the theorem. Let Δ be a diagram in $\mathbf{A}$ with objects Δ_n at nodes $n \in nodes(\Delta)$. Let the colimit of Δ be $\xi_n : \Delta_n \to a$. We show that $F(\xi_n) : F(\Delta_n) \to F(a)$ is a colimit of $F(\Delta)$ in $\mathbf{B}$. Let $\nu_n : F(\Delta_n) \to b$ be a cocone on $F(\Delta)$ then $G(\nu_n)\eta(\Delta_n) : \Delta_n \to G(b)$ is a cocone on Δ and so there is a unique $v : a \to G(b)$ such that the square below commutes for all $n \in nodes(\Delta)$.

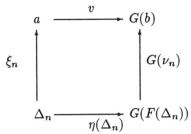

Finally, by the definition of adjunctions, there is a unique $v^\# = \epsilon(b)F(v)$ such that the following commutes.

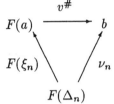

This establishes the cocontinuity of the left adjoint. $\square$

The proof constructs cocontinuous functors from adjunctions. We express this construction in the program below, where the cocontinuity is represented as a preservation function from colimits to colimits:

```
fun adj_cocontinuous(A as adjunction(F,G,eta,epsilon))=
      { the preservation of colimiting cocones }
   let val preserve =
     fn (c,u) =>
         { the cocone is the application of F }
      let val result_cocone = apply_Fun_CoCone(F,c)
          { the universal part }
        val universal =
          fn c1 =>
           let val c2 =
             new_cocone(domain F)
                ( apply_Nat_Diag(eta,base(c)),
                  apply_Fun_CoCone(G,c1))
             val v = co_apex_arrow(u c2)
             val (_,v_sharp,_) =
                sharp(A) ofo (co_apex(c),v,co_apex(c1))
           in cocone_arrow(result_cocone,v_sharp,c1) end
       in (result_cocone,universal) end
    in cocontinuous_functor(F,preserve) end
```

The function **new_cocone** takes a cocone $\xi_n : \Delta'_n \to a$ on diagram Δ' and a diagram arrow $\delta : \Delta \to \Delta'$ and returns the cocone $\xi_n\delta_n : \Delta_n \to a$.

Duality, which allows us to state that the right adjoint is continuous, can be coded using the above program and conversion functions:

```
fun adj_continuous(adj) =
        dual_CoCon_Fun(adj_cocontinuous(dual_Adj(adj)))
```

The dual of an adjunction is defined above. The fact that the dual of a cocontinuous functor is a continuous functor can be similarly encoded.

6.5   Free algebras

In Section 6.3.3 we described a construction of free algebras as term algebras. The construction is iterative, building terms of increasing depth. There is a categorical generalization of this construction which we now present. The generalization allows us to move away from algebras whose carriers are in **Set** to other categories. To do so we replace operator domains by functors and represent algebras as arrows corresponding to the evaluation of terms.

An algebra of a functor $F : \mathbf{C} \to \mathbf{C}$ is an object a of $\mathbf{C}$ and an arrow $f : F(a) \to a$. An arrow between algebras $h : (a, f) \to (a', f')$ is an arrow $h : a \to a'$ such that the following square commutes.

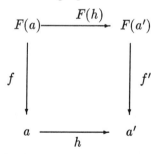

The category of F-algebras with these arrows we denote by $(F : \mathbf{C})$. The carrier a of algebra (a, f) determines a functor $U : (F : \mathbf{C}) \to \mathbf{C}$. In the next section we show how to construct a left adjoint to U, i.e. free F-algebras.

Example

An operator domain Ω determines a functor $F : \mathbf{Set} \to \mathbf{Set}$ defined by

$$F(X) = \{\rho(x_1, x_2, \ldots, x_n) | n \in \mathcal{N}, \rho \in \Omega_n, x_i \in X, 1 \le i \le n\}.$$

Thus $F(X)$ is the set of terms of depth 1 in X. An Ω-algebra (A, α) determines a function $f : F(A) \to A$ by

$$f(\rho(a_1, a_2, \ldots, a_n)) = \alpha_\rho(a_1, a_2, \ldots, a_n).$$

There is another functor associated with an operator domain $F' : \mathbf{Set} \to \mathbf{Set}$, $F'(X) = X \cup F(X)$, terms of depth at most 1 in X. There is an inclusion $X \to F'(X)$ and again an Ω-algebra (A, α) determines a function $F'(A) \to A$. The term algebra $T_\Omega(X)$ (see Section 6.3.3) is a free F-algebra on set X (and also a free F'-algebra).

6.5.1 Constructing free algebras

We now look at some iterative constructions of free F-algebras. There has been extensive study of such constructions, including transfinite iterations, and their application, for instance to constructing colimits of algebras. The following publications illustrate this material: [Barr 70], [Schubert 72], [Dubuc 74], [Manes 76], [Adámek, Trnkova 78], [Adámek 77,78], [Adámek, Koubek 80] and [Barr, Wells 85].

There are various constructions of free F-algebras involving countable coproducts or ω-colimits, that is, colimits of diagrams (called ω-chains) of the form:

$$a_0 \xrightarrow{f_0} a_1 \xrightarrow{f_1} a_2 \xrightarrow{f_2} \cdots a_n \xrightarrow{f_n} a_{n+1} \cdots$$

Their applicability depends upon preservation properties of the functor F. In this section we consider a construction using countable coproducts and use it to compute the transitive closure of a graph. Later, as exercises, we present two ω-colimit constructions of free algebras.

Theorem 5 *If* $\mathbf{C}$ *has countable coproducts and* $F : \mathbf{C} \to \mathbf{C}$ *preserves them, then free F-algebras exist. The free algebra on a is constructed as the coproduct:*

$$a + F(a) + F^2(a) + F^3(a) + \cdots$$

Proof Let $\xi_n : F^n(a) \to T(a)$ $(n \geq 0)$ be the coproduct

$$a + F(a) + F^2(a) + F^3(a) + \cdots$$

Then T is a functor. The action of T on arrows is defined by universality of the coproduct. Since F preserves countable coproducts $F(\xi_n) : F^{n+1}(a) \to F(T(a))$ is the coproduct

$$F(a) + F^2(a) + F^3(a) + \cdots$$

But $\xi_n : F^n(a) \to T(a), (n \geq 1)$ is a cocone on these objects, so there is a unique arrow $f : F(T(a)) \to T(a)$ such that, for $n \geq 1$, $fF(\xi_n) = \xi_n$. This is the free algebra. There is an arrow $\xi_0 : a \to T(a)$ and the universality arises from that of the coproduct.

6.5.2   A program

We now program this construction of free algebras and use it to compute the transitive closure of graphs.

The iterative construction can be encoded as it stands. To apply the construction, we need categories with countable coproducts. Now, the category **FinSet** does not have all countable coproducts and, although **Set** does, coproducts of finite sets may well be infinite. The case we deal with here is when the iteration terminates. This seemingly trivial case is a categorical version of program iteration, which we code as a recursive program in ML. Termination means that the sequence of objects eventually becomes constantly the initial object. Notice that, for chains generated by functors preserving the initial object, the first occurrence of an initial object guarantees that the sequence remains constant thereafter. This makes termination checkable.

Iterative constructions of free algebras, like that above, require the functor F to preserve certain colimits. In programming this, it is not sufficient that we know the preservation property holds, we need to express it in the program for it provides the algebraic structure of the free algebra. The preservation properties of F depend upon how F is constructed in terms of limits and colimits. These preservation properties are intricate to establish and hold for a restricted range of categories. Rarely are general theorems available. For constructing free algebras of equational theories over **Set**, the relevant preservation property is the commutation of finite limits with filtered colimits in **Set**. Such specialized constructions may be realized as, albeit lengthy, programs.

Alternatives are available. We may operate under the assumption that the relevant colimits are preserved and compute two colimits, that of the original diagram and that of the result of applying F to the diagram. The latter is used in place of the image of the colimit under F. This necessarily involves extra computation and is the penalty for avoiding programming the proof of preservation. In the case of colimits of ω-chains of arrows, termination means that the arrows eventually all become isomorphisms. Because functors preserve isomorphisms, the preservation of colimits of terminating ω-chains is simply expressed.

We now program the construction. We say°that the countable co-product of objects a_i, $i \in \omega$, *terminates* if there is an $n \in \omega$ such that $\forall i \geq n$. a_i is an initial object. In this case the colimit is a finite coproduct. When the functor F preserves initial objects, it suffices for termination of $F^i(a)$, $i \in \omega$, that $F^n(a)$ is the initial object for some $n \in \omega$.

The following program accumulates the countable coproduct and tests for termination, under the assumption that F preserves initial objects, by testing each iterate for isomorphism with the initial object. We thus need categories in which isomorphisms are recognizable and inverses are constructible. We give examples of these in the next section.

```
datatype ('o,'a)IsoCat =
    iso_cat of ('o,'a)Cat * ('a -> bool) * ('a -> 'a)
type ('o,'a)Countable_Coproduct =
    (num ->'o) * 'o * (num->'a) * ('o * (num->'a) -> 'a)

fun iterated_coproduct(cC,iso_cat(C,is_iso,_))(F)(a) =
  let fun accumulate(Fna,sum) =
        if is_iso(initial_arrow(cC)(Fna)) then sum
        else accumulate(F ofo Fna, add(cC)(n,Fna,sum)) in
  accumulate(a,initial_coproduct(cC)) end
```

The function initial_coproduct is the coproduct of the chain of initial objects and add adds another iterate to a repeated coproduct using the binary coproduct in the finitely cocomplete category cC.

The free algebra is constructed from the universality of the coproduct and the preservation of coproducts. The latter may be replaced by a further computation of colimits as discussed above.

```
fun free_algebra(cC,isoC)(F)(a) =
  let val ((_,Ta,xi),univ) =
            iterated_coproduct(cC,isoC)(F)(a)
      val (_,Funiv) =
            iterated_coproduct(cC,isoC)(F)(F ofo a)
      val result_alg = (Ta,Funiv(Ta,fn n => xi(succ n)))
      val universal =
       fn ((b,epsilon'),f) =>
       let fun nu(zero) = f
             | nu(succ n) =
                 compose(cat_of cC)(epsilon',F ofa nu(n))
       in (result_alg,univ(b,nu),(b,epsilon')) end in
  (result_alg,universal) end
```

6.5.3   An example: transitive closure

We apply this construction of free F-algebras to compute the transitive closure of a graph. The functor F is constructed using limits and colimits, which means that we can assemble a program for transitive closure from pieces already available.

A path in a graph is a non-empty sequence of edges $e_1, e_2, \ldots, e_n$, $n > 0$ such that $t(e_i) = s(e_{i+1})$, $0 < i < n$, where s and t are the source and target, respectively, of the edges. The transitive closure of the graph has the same nodes, but has paths as edges. A path ending at a node may be composed, by concatenation, with a path starting at the node and this composition is associative. Also there are source and target operations on paths yielding the start and end nodes. Under these operations, the transitive closure of a graph forms an algebra. This is an algebra on graphs rather than sets. In fact, transitive closure is the free algebra on a graph. All this is well known: if we include identities at each node (empty paths), the transitive closure becomes the path category, the free small category on a graph.

Recall that graphs in a category $\mathbf{C}$ are parallel pairs of arrows. The transitive closure is a graph on the same nodes as the original graph. We introduce a category of graphs on a fixed set of nodes as follows. Define the category $\mathbf{Graph}_a(\mathbf{C})$ to have as objects parallel pairs in $\mathbf{C}$ whose target is a and, as arrows, triangles which serially commute. This is a 'double' version of a slice category.

```
datatype ('o,'a)Parallel_Pair = pp of 'a * 'a
datatype ('o,'a)Parallel_Pair_Arrow = pp_arrow of
        ('o,'a)Parallel_Pair * 'a * ('o,'a)Parallel_Pair
fun Graph(C)(a) =
 cat(fn pp_arrow(s,_,_) => s,
    fn pp_arrow(_,_,t) => t,
    fn pp(f,g) =>
          pp_arrow( pp(f,g),
                    identity(C)(source(C)(f)),
                    pp(f,g) ),
    fn (pp_arrow(_,f',q'),pp_arrow(p,f,q)) =>
                pp_arrow(p,compose(C)(f',f),q'))
```

If $\mathbf{C}$ is (finitely) cocomplete then so is $\mathbf{Graph}_a(\mathbf{C})$ and the construction yields a program.

Consider for the moment graphs in $\mathbf{FinSet}$. Define a binary operation $*$ taking graphs G and G' on the same set of nodes to a graph on these

nodes and having, as edges, composable pairs of edges, the first from G, the second from G':
$$edges(G * G') =$$

$$\{\langle e_1, e_2 \rangle : e_1 \in edges(G), e_2 \in edges(G'), t_G(e_1) = s_{G'}(e_2)\}.$$

This is the composition of graphs considered as relations and so may be defined in any category with pullbacks: Consider graphs $G = \langle s, t \rangle :$ $c \to a$ and $G' = \langle s', t' \rangle : c' \to a$. Let the following be a pullback square:

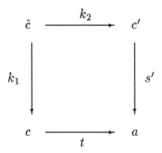

Then $G * G' = \langle sk_1, t'k_2 \rangle : \hat{c} \to a$. This operation extends to a functor. The action of $*$ on arrows is defined in terms of the universality of the pullback as described in the program below:

```
fun edge_composition(1C as complete_cat(C,_))
                (G as pp(s,t))(G' as pp(s',t')) =
        let val ((k1,k2),pb_univ) = pullback(1C)(t,s') in
        pp(compose(C)(s,k1),compose(C)(t',k2)) end

fun action_on_arrows(1C as complete_cat(C,_))
        (pp_arrow(G as pp(s,t),f,G' as pp(s',t')))
        (pp_arrow(H as pp(u,v),g,H' as pp(u',v'))) =
    let val ((k1,k2),pb_univ)   = pullback(1C)(t,s')
        val ((k1',k2'),pb_univ') = pullback(1C)(v,u')
        val u =
        pb_univ'(compose(C)(f,k1),compose(C)(g,k2)) in
    pp_arrow( edge_composition(1C)(G)(H),
             u,
             edge_composition(1C)(G')(H')) end
```

```
fun star(1C as complete_cat(C,_))(a) =
    bifunctor( Graph(C)(a),Graph(C)(a),
               edge_composition(1C),
               action_on_arrows(1C),
               Graph(C)(a) )
```

A bifunctor is a binary functor, i.e. a functor from a product of two categories.

Theorem 5 may be used to accumulate the transitive closure as a countable coproduct by noting that, for graphs in **FinSet**, the functor $\Sigma_G : H \mapsto G * H$ preserves initial objects and (finite and) countable coproducts. The free Σ_G algebra on G is a graph $T(G)$ and a graph arrow $G * T(G) \to T(G)$. The graph $T(G)$ is the transitive closure of G, the arrow corresponds to the operation of extending a path by concatenating an edge onto its front. The freeness of this construction ensures that edges in $T(G)$ are indeed paths of edges.

To program this construction of transitive closure, we must first consider categories with recognizable isomorphisms and explicit inverses to isomorphisms. Isomorphisms in **FinSet** are functions whose source, target and image all have the same cardinality:

```
val isoFinSet =
    let fun is_iso(fa as set_arrow(a,f,b)) =
        (cardinal(a)=cardinal(b)) andalso
              (cardinal(b)=cardinal(image fa))
        fun invert(set_arrow(a,f,b)) =
        set_arrow( b,
                      fn x =>
                        let val S =
                        filter(fn z => (f(z)=x))(a)
                            val (y,_) =
                        singleton_split(S) in y end,
                      a ) in
    iso_cat(FinSet,is_iso,invert) end
```

There is an inheritance of isomorphism structure from a category $\mathbf{C}$ to the category $\mathbf{Graph}_a(\mathbf{C})$.

```
fun iso_Graph(iso_cat(C,is_iso,invert))(a) =
    iso_cat( Graph(C)(a),
              fn pp_arrow(_,f,_) => is_iso(f),
              fn pp_arrow(s,f,t) =>
                     pp_arrow(t,invert(f),s) )
```

Now we have the ingredients of a transitive closure program:

```
fun transitive_close(G) =
    let val N = pp_source(FinSet)(G) in
    free_algebra( cocomplete_Graph(cocomplete_FinSet)(N),
                  iso_Graph(isoFinSet)(N) )
                 (as_functor(star(complete_FinSet),G))(N)
    end
```

Notice what has gone into this program: colimits (coproducts) in **FinSet** and the inheritance by graphs, recognition of isomorphisms and limits (pullbacks) in **FinSet**. This may seem excessive for the transitive closure of graphs but notice that we calculate not only the transitively closed graph but also its algebraic structure and its universality. Moreover, there is an implied generality of this code as we may release it from graphs over **FinSet** to more general graphs, for instance, with structure on the node or edge sets.

This algorithm is the repeated accumulation of pairs of paths and so has complexity $O(N^3 \log N)$ where N is the number of nodes. Faster algorithms exist, see [Warshall 62] for an $O(N^3)$ algorithm. A different abstract approach to transitive closure uses semirings (see e.g. [Lehmann 77]).

6.5.4 Other constructions of free algebras

In a series of exercises we present two constructions of free algebras in terms of colimits of ω-chains. An ω-chain is a diagram of the form:

$$a_0 \xrightarrow{\ f_0\ } a_1 \xrightarrow{\ f_1\ } a_2 \xrightarrow{\ f_2\ } \cdots \ a_n \xrightarrow{\ f_n\ } a_{n+1} \ \cdots$$

We let the arrows f_n, $n \in \omega$ represent ω-chains:

```
type ('o,'a)w_Chain = (num -> 'a)
```

An ω-cocone is a cocone on an ω-chain and an ω-colimit is a colimiting ω-cocone. These are represented as types following the treatment of finite colimits.

```
datatype ('o,'a)w_CoCone =
   w_cocone of 'o * ('o,'a)w_Chain * (num -> 'a)
datatype ('o,'a)w_CoCone_Arrow  =
   w_cocone_arrow of ('o,'a)w_CoCone*'a*('o,'a)w_CoCone
type ('o,'a)w_Colimiting_CoCone =
   (('o,'a)w_CoCone *
       (('o,'a)w_CoCone -> ('o,'a)w_CoCone_Arrow))
type ('o,'a)w_Colimit =
       ('o,'a)w_Chain -> ('o,'a)w_Colimiting_CoCone
datatype ('o,'a)w_CoComplete_Cat  =
       w_cocomplete_cat of ('o,'a)Cat * ('o,'a)w_Colimit
```

Exercise 7. ω-colimits of terminating chains.

A chain $f_n : a_n \to a_{n+1}$, $n \in \omega$ is said to *terminate* iff there is an $N \in \omega$ such that for all $n \geq N$ the arrows f_n are isomorphisms (i.e. are invertible). Such chains trivially have colimits. Any a_n, $n \geq N$ is a colimiting object.

Write a program to construct colimits of terminating ω-chains in any category with recognizable isomorphisms and explicit inverses to isomorphisms. Assume that the first occurrence of an isomorphism in the chain guarantees that all succeeding arrows are isomorphisms. This will be the case for the free algebra constructions which we present below.

As a help, we give the function which finds the first occurrence of an isomorphism:

```
fun fixed_point(isoC as iso_cat(C,iso,inverse))
              (w_chain)(n) =
       if iso(w_chain n)
       then n else fixed_point(isoC)(w_chain)(n+1)
```

Exercise 8. The following theorem constructs free F-algebras as ω-colimits.

Theorem 6 *Let* **C** *have colimits of countable chains (ω-colimits) and an initial object 0. Let* $F : \mathbf{C} \to \mathbf{C}$ *preserve ω-colimits. Then, the initial F-algebra exists and is constructed as the colimit of the chain:*

$$0 \xrightarrow{\ f\ } F(0) \xrightarrow{\ F(f)\ } F^2(0) \xrightarrow{\ F^2(f)\ } \cdots \ F^n(0) \xrightarrow{\ F^n(f)\ } F^{n+1}(0) \cdots$$

where $f : 0 \to 0F$ is the unique such arrow.

In the case that **C** has binary coproducts, the free algebra on a is obtained by replacing F by $a + F$ in the above construction, where a stands for the constant functor returning the object a.

Prove this and encode the construction.

Exercise 9. Here is a construction of free algebras using an iterated pushout:

Theorem 7 *Let* **C** *have finite colimits and ω-colimits. Let F : **C** $\to$ **C** *preserve ω-colimits. Then free F-algebras exist. The free algebra on a, $F(a_\infty) \to a_\infty$, is constructed as the colimit of the following chain of pushout squares,*

$$
\begin{array}{ccccccccc}
0 & \xrightarrow{\ f\ } & F(a_0) & \xrightarrow{\ F(f_1)\ } & F(a_1) & \xrightarrow{\ F(f_2)\ } & F(a_2) & \xrightarrow{\ F(f_3)\ } & \cdots \\
\downarrow{\scriptstyle g} & & \downarrow{\scriptstyle g_1} & & \downarrow{\scriptstyle g_2} & & \downarrow{\scriptstyle g_3} & & \\
a_0 & \xrightarrow{\ f_1\ } & a_1 & \xrightarrow{\ f_2\ } & a_2 & \xrightarrow{\ f_3\ } & \cdots & &
\end{array}
$$

where $a_0 = a$ and f and g are the unique arrows from the initial object.

Prove this theorem.

The function finding when such a chain of pushouts terminates is given below. Use this to write a program constructing free algebras as in the theorem.

```
fun fixpoint(cC,isoC as iso_cat(C,is_iso,_))(F)(a) =
   let fun accumulate(f_n,g_n) =
        let val ((f_succn,g_succn),_) =
                    pushout(cC)(g_n,f_n) in
        if is_iso(f_succn) then source(C)(f_n) else
        accumulate(F ofa f_succn,g_succn) end in
   accumulate( initial_arrow(cC)(a),
               initial_arrow(cC)(F ofo a)) end
```

This construction may be used to compute colimits of algebras and free algebras of equational theories. In the latter case, the equations may be imposed at each step of the iteration using appropriate coequalizers. More about the iterated pushout construction may be found in [Adámek, Trnkova 78], [Adámek 77,78], [Adámek, Koubek 80] and [Barr, Wells 85].

6.6   Exercises

Exercise 10. Arrows of adjunctions can be defined to form a category of adjunctions: Let **A** and **B** be categories. A pair of natural transformations $(\sigma : F \to F', \tau : G' \to G)$ from adjunction $\langle F, G, \eta, \epsilon \rangle : \mathbf{B} \to \mathbf{A}$ to adjunction $\langle F', G', \eta', \epsilon' \rangle : \mathbf{B} \to \mathbf{A}$ is said to be *conjugate* if the diagram below commutes:

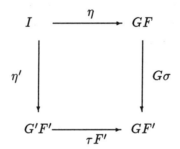

In fact, each natural transformation determines the other.

Under the composition $(\sigma, \tau).(\sigma', \tau') = (\sigma.\sigma', \tau'.\tau)$, adjunctions, with conjugate pairs of natural transformations as arrows, form a category **Adj(B, A)**. Give a computational representation of the category of adjunctions.

Exercise 11. Adjunctions may be composed. Given $\langle F, G, \eta, \epsilon \rangle : \mathbf{A} \to \mathbf{B}$ and $\langle F', G', \eta', \epsilon' \rangle : \mathbf{B} \to \mathbf{C}$, their composition is defined to be the adjunction $\langle FF', G'G, \eta.F\eta'G, G'\epsilon F'.\epsilon' \rangle : \mathbf{A} \to \mathbf{C}$.

Exercise 12. Show that if **C** is complete so is the category of algebras $(F : \mathbf{C})$. Program this construction of finite limits.

Constructions of colimits of algebras are not so straightforward. Iterative constructions are available as mentioned in Exercise 9.

Exercise* 13. Minimal realization by automata Goguen [1973], as well as Arbib and Manes [1974,75a], show that the minimal realization of a behaviour by a finite automaton is right adjoint to a behaviour functor. Write a program to compute minimal realizations as an adjunction. A reference for the construction is [Hopcroft, Ullman 79].

Exercise* 14. Monads Free algebras find a setting in the theory of monads. Monads are an alternative formulation of adjunctions and provide a description of equational theories in terms of functors and natural transformations. Suitable references are [Mac Lane 71], [Manes 76] and [Barr, Wells 86]. Many of the constructions involving monads may be expressed as programs. Try this.

Alagić [1975] has shown how top-down and bottom-up tree processing may be given a categorical setting using monads. Can you use these ideas to write general tree-processing routines?

Exercise* 15. Consider the category of recursively enumerable (r.e.) sets with recursive functions as arrows. Which limits and colimits exist in this category?

Lazy lists in ML, otherwise called streams, may be used to represent r.e. sets and hence limits and colimits that exist may be encoded.

Exercise* 16. Constructing adjunctions via colimits The fact that left adjoints are cocontinuous may be used to construct adjunctions as follows.

Let $F : \mathbf{A} \to \mathbf{B}$ be a cocontinuous functor. To compute the action of F on an object a of $\mathbf{A}$, express a as a colimit of a diagram Δ and assume, recursively, that we can compute $F(\Delta)$. Then, by cocontinuity, $F(a)$ is the colimiting object of $F(\Delta)$ in $\mathbf{B}$. Notice that in computing $F(\Delta)$ we need to (recursively) compute the action of F on arrows as well as objects. This requires careful analysis since the expression of objects and arrows as colimiting cocones and arrows of colimiting cocones respectively is not, in general, functorial. Conditions can be imposed to make this scheme work. In particular, if F is a left adjoint then, assuming the recursion terminates, we construct adjunctions using colimits. This is the substance of the following theorem:

Theorem 8 *Consider a diagram of functors:*

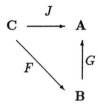

Let $J : \mathbf{C} \to \mathbf{A}$ satisfy the following: For each object a of $\mathbf{A}$, there is a diagram Δ in $\mathbf{C}$ such that a is a colimit of $J(\Delta)$.

Suppose $\mathbf{B}$ is cocomplete and there is a natural transformation $\eta : J \to GF$ such that, for each object c of $\mathbf{C}$, $(F(c), \eta(c))$ is free on $J(c)$ with respect to G, then F extends (necessarily uniquely) to a left adjoint $F^* : \mathbf{A} \to \mathbf{B}$ of G such that $F^*J = F$.

Note: Such an F and η is equivalent to, for each c in $\mathbf{C}$, a (canonical) universal arrow from $J(c)$ to G.

Prove this theorem and show how it leads to recursive programs for computing adjunctions assuming that the decomposition into diagrams terminates.

Show how Warshall's algorithm for the transitive closure of a graph can be considered to be an example of this construction of adjunctions. Warshall's algorithms is described in [Warshall 62] and [Aho, Hopcroft, Ullman 74].

Chapter 7

Toposes

This is the final chapter in which we consider basic category theory and its computational representation. We shall take a brief look at categorical logic, considering cartesian closed categories and toposes.

Cartesian closed categories are categories in which analogues of function spaces, called exponentials, are defined. There is an exact correspondence between cartesian closed categories and certain λ-calculi, see [Lambek 69,80] and [Lambek, Scott 86]. This correspondence provides an algebraic treatment of models of λ-calculi. Moreover, it translates functional languages with variables into a variable-free combinator language. Curien [1986] uses the combinators of cartesian closed categories to define an abstract machine for implementing functional languages.

Elementary toposes were introduced by Lawvere and Tierney. Toposes are cartesian closed categories which have an extra structure allowing the definition of an internal logic which is, in general, intuitionistic rather than classical. Just as certain λ-calculi correspond to cartesian closed categories so theories in higher order intuitionistic logic correspond to toposes. The logical operations are expressed in terms of limits, colimits and exponentials and can be realized as programs. Originally, toposes were introduced in an attempt to characterize the category **Set** in an entirely arrow-theoretic fashion. So interpreted, toposes provide categorical primitives for manipulating sets so that programs involving sets may be assembled from the operations in a topos.

In this chapter, the treatment is fairly brief. We define cartesian closed categories and toposes, giving computational representations and some example computations involving these concepts. This is all preliminary to presenting programs to compute the internal logic in toposes. These programs are run on an example of a topos to provide truth-tables for a three-valued logic.

For those wishing to know more about categorical logic and toposes, useful references are [Freyd 72], [Goldblatt 79], [Johnstone 77], [Seely 83] [Barr, Wells 85], [Lambek, Scott 86] and the introductory paper [Mac Lane 75].

7.1   Cartesian closed categories

It is somewhat remarkable that function spaces can be universally characterized:

Definition 23 *Let* **C** *be a category with binary products. If a and b are objects of* **C***, an* exponential *from a to b is an object b^a and an arrow $e : a \times b^a \to b$ such that for any object c and arrow $f : a \times c \to b$, there is a unique arrow $f^\# : c \to b^a$ such that the triangle below commutes.*

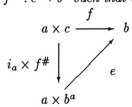

Definition 24 *A category which is finitely complete and has exponentials for any pair of objects is called* cartesian closed.

Some authors require only finite products rather than all finite limits. Notice that the definition of exponentials is that of an adjunction: the functor $(\_)^a$ is right adjoint to the functor $a \times \_$.

Just as in previous chapters we represented universally defined concepts as functional types in ML, so here we represent exponentials and hence declare a type for cartesian closed categories:

```
type ('o,'a)Exponential =
        ('o*'o) -> ( ('o*'a) * (('o*'a) -> 'a) )
datatype ('o,'a)Cartesian_Closed_Cat  =
    cartesian_closed_cat of
        ('o,'a)Cat * ('o,'a)Limit * ('o,'a)Exponential
```

7.1.1   An example: the category of finite sets

The categories **FinSet** and **Set** are both cartesian closed. The exponential b^a is the set of all functions from a to b, the function $e : a \times b^a \to b$ is application `apply(x,f) = f(x)` and the universal property is the so-called 'curry' operation converting a binary operation into repeated application:

$$\texttt{curry(f) = fn x => fn y => f(y,x)}$$

To program the exponential in **FinSet**, we represent functions by their graphs (sets of pairs) and then compute the set of all functions between two finite sets. The latter calculation is based on the following isomorphisms which hold in any cartesian closed category with initial object 0 and binary coproducts:

$$b^0 \cong 1$$

$$b^1 \cong b$$

$$b^{a+a'} \cong b^a \times b^{a'}$$

These may be verified using elementary arguments and the definition of exponentials. Alternatively, the first and third isomorphisms arise from the fact that the functor $b^{(\text{-})} : dual(\mathbf{C}) \to \mathbf{C}$ is right adjoint to itself and so carries colimits to limits.

These isomorphisms supply a recursive algorithm for calculating the exponential in **FinSet** as follows: The finite set a in b^a is either empty, in which case we have $b^0 \cong 1$, or is a singleton set, in which case $b^1 \cong b$ or it can be expressed non-trivially as a coproduct, in which case we have $b^{a+a'} \cong b^a \times b^{a'}$.

Again, we need to ensure that the elements of the sets are closed under the operations on them, in this case that of forming lists. We extend the type of these elements, which already are closed under operations like pairing:

```
datatype 'a Tag = ... | tuple of ('a*'a)list
```

The function space b^a is computed by case analysis on set a:

```
fun function_space(a,b) =
        if cardinal(a) = 0 then singleton(tuple [])
    else if cardinal(a) = 1 then
            let val x = element_of(a) in
```

```
                 mapset((fn y => tuple([x,y)])),b) end
        else let val (a1,a2) = split(a)
                 val prod =
                    cartesian_prod( function_space(a1,b),
                                    function_space(a2,b)) in
        mapset(fn (tuple(x),tuple(y))=>tuple(append(x,y))),
                 prod) end
```

The exponential is the function space together with an evaluation arrow
$e : a \times b^a \rightarrow b$.

```
    fun exponential(a,b) =
     let val fs = function_space(a,b)
         val ((fs_X_a,proj_fs,proj_a),_) =
                   product(complete_FinSet)(fs,a)
         val eval =
           set_arrow(fs_X_a,
                         fn pr => let val s = proj_fs OF pr
                                      and x = proj_a OF pr
                                      in list_to_fn(s)(x) end,
                         b )
     in (fs,eval) end
```

The description is completed by defining the universal property as a
function taking a pair $(c, f : a \times c \rightarrow b)$ and returning an arrow $f^{\#} : c \rightarrow$
b^a:

```
    fun set_exp(a,b) =
     ( exponential(a,b),
       fn (c,f) =>
          let val (_,univ) = product(complete_FinSet)(a,c)
              val tt_obj as (tt,tt_arrow) =
                      terminal_obj(complete_FinSet)
              fun constant_arrow(x,d) =
                      set_arrow(terminal_obj,constant(x),d)
              val comp = compose(FinSet)
          in set_arrow( c,
                        fn y =>
                          let fun fsharp(z) =
                              comp(e,
                                   univ(tt_obj,
                                        constant_arrow(z,a),
```

```
                              constant_arrow(y,c))
                   OF ttrue
               in fn_to_list(fsharp,a) end,
           function_space(a,b) ) end )
```

The cartesian closed category of finite sets is the category **FinSet** together with limits and exponentials:

```
val cc_FinSet =
      let val complete_cat(_,lim) = complete_FinSet in
      cartesian_closed_cat(FinSet,lim,set_exp) end
```

7.2 Toposes

A topos is a cartesian closed category which has an object, called a subobject classifier, to represent the truth-values. In the topos **Set** this is a two element set (for the truth-values 'true' and 'false'). Other toposes allow more exotic truth-values, including logics of undefinedness and temporal logics. Internal logics in toposes are intuitionistic. Indeed, there is a exact correspondence between toposes and theories in higher order intuitionistic logic, see [Lambek, Scott 86].

The following definition captures the required behaviour of truth-values in arrow-theoretic terms:

Definition 25 *Let* **C** *be a category with a terminal object,* 1. *A subobject classifier in* **C** *is an object of* **C**, Ω, *together with an arrow* $t : 1 \rightarrow \Omega$, *which satisfy the following:*

For each monic $m : a \rightarrow c$ *there is a unique* $\chi_m : c \rightarrow \Omega$, *called the character of* m, *such that the following is a pullback square.*

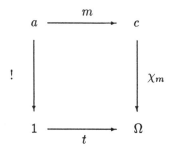

We use the notation $! : a \to 1$ for the unique arrow into the terminal object. Recall that an arrow $m : b \to c$ is a monic iff for all arrows $f, g : a \to b$, $mf = mg \Rightarrow f = g$.

In **Set**, Ω is a two-element set, say, $\{tt, f\!f\}$ and the arrow $t : 1 \to \Omega$ selects the element tt. A monic (a 1-1 function) $m : a \to c$ divides c into the elements that are in the image under m and those that are not. The character of m maps image elements to tt and other elements to $f\!f$. It is readily verified that this does indeed make the above square a pullback.

Definition 26 *A category is a* topos *if*

- *It is finitely bicomplete (i.e. complete and cocomplete),*

- *It has exponentials,*

- *It has a subobject classifier.*

It can be shown that cocompleteness follows from the other axioms, so we may define a topos as a cartesian closed category with a subobject classifier.

In a topos, the object Ω is called the *object of truth values*, the arrow $t : 1 \to \Omega$ is *truth* and for any object a the composite $a \xrightarrow{\;!\;} 1 \xrightarrow{\;t\;} \Omega$ is called *true* on a. The *name* of an arrow $f : a \to b$ in any cartesian closed category is the unique arrow $\hat{f} : 1 \to b^a$ determined by the universality of the exponential b^a.

Turning to the computational representation of toposes, we see that the definition of a topos involves monic arrows. Monics are not simply arrows with extra computational structure. Any treatment of monics requires an equality on arrows – either a decidable equality or a proof system for establishing equality. However, constructions involving toposes do not make reference to the monic nature of arrows which rather is part of the correctness argument for the constructions.

We proceed to represent toposes and give examples of constructions within them including that of the internal logic:

```
type ('o,'a)PullBack_Square =
      ('a*'a*'a*'a) * (('a*'a) -> 'a)
type ('o,'a)Subobject_Classifier =
      ('o*'a) * ('a -> ('o,'a)PullBack_Square)
```

```
datatype ('o,'a)Topos =
            topos of  ('o,'a)Cat *
                      ('o,'a)Limit *
                      ('o,'a)Colimit *
                      ('o,'a)Exponential *
                      ('o,'a)Subobject_Classifier
```

Here are some useful functions associated with toposes:

```
fun true_ (T as topos(C,lim,_,_,((_,truth),_)))(a) =
    let val cC = complete_cat (C,lim) in
      compose(C)(truth,terminal_arrow(cC)(a)) end

fun character(topos(_,_,_,_,(_,sc_univ)))(m)  =
      let val ((_,chi,_,_),_) = sc_univ(m) in chi end

fun name (topos(C,lim,_,exp,_))(f) =
    let val cC = complete_cat(C,lim)
        val t_obj = terminal_obj cC
        val a = source(C)(f) and b = target(C)(f)
        val pr_a_f = let val ((_,pr_a,_),_) =
                              product cC (t_obj,a)
                     in compose(C)(f,pr_a) end
        val (_,exp_adj) = exp(a,b)
    in exp_adj (t_obj,pr_a_f) end
```

7.2.1   An example: the topos of finite sets

The categories **Set** and **FinSet** are both toposes. We describe the topos **FinSet** as a value in ML. Since we already have programmed the cartesian closed category of finite sets and also finite colimits in **FinSet**, all that remains is the subobject classifier. This was described in the previous section and is expressed in the following program:

```
val subobject_classifier =
    let val 1C as complete_cat(C,_) = complete_FinSet
        val truvals = {ttrue,ffalse}
        val truth =
          set_arrow({ttrue},fn x=>ttrue,truvals)
        val chi =
          fn m =>
```

```
            set_arrow(target(C)(m),
                      fn z =>
                      if member(z,(image_set m))
                         then ttrue else ffalse,
                      truvals )
        in ( (truvals,truth),
          fn m =>
            let val t =
              terminal_arrow(1C)(source(C)(m)) in
            ((truth,chi(m),t,m),
              fn (f,g) =>
                compose(C)(right_inverse(m),g)) end) end
      val topos_of_sets =
        let val cartesian_closed_cat(C,lim,exp) = cc_FinSet
            val cocomplete_cat(_,colim) = cocomplete_FinSet
        in topos(C,lim,colim,exp,subobject_classifier) end
```

7.2.2 Computing in a topos

Constructions within a topos can be programmed using finite limits and colimits, exponentials and the subobject classifier.

As an example we consider the concept of power objects. This is a universal characterization of the powerset construction in the toposes **Set** and **FinSet**.

Definition 27 *Let* **C** *be a category with binary products and a an object of* **C**, *the* power object *of a is an object of* **C**, $P(a)$, *together with an object d and a monic* $p : d \to a \times P(a)$, *such that, for any object b and monic* $m : r \to a \times b$, *there is a unique arrow* $f : b \to P(a)$ *and a (necessarily unique) arrow* $g : r \to d$ *to make the square below a pullback square:*

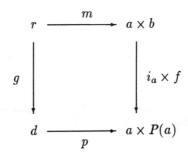

Proposition 10 *In a topos, every object has a power object.*

Proof The idea behind the construction is that in **Set** we can represent the powerset $P(a)$ by the set of functions from a to the two-element set of truth-values Ω. To lift this to an arbitrary topos, we need only define the requisite arrows and check their universality as follows.

For an object a of topos **T** with subobject classifier $t : 1 \to \Omega$ and exponential $e : a \times \Omega^a \to \Omega$, define $P(a) = \Omega^a$ and $p : c \to a \times P(a)$ by the pullback square:

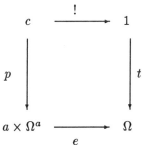

The universal property of $(P(a), p)$ can be established from that of the subobject classifier, the exponential and the above pullback. $\square$

This construction is given explicitly in the ML function below.

```
type ('o,'a)Power =
  'o -> ('o*'a*('a -> ('a*('o,'a)PullBack_Square)))

fun power(topos(C,lim,colim,exp,sc)) =
  let val ((truvals,truth),character_square) = sc
      val 1C = complete_cat(C,lim) in
  fn a =>
     let val ((P_of_a,eval),exp_adjoint) =
                         exp(a,truvals)
         val ((h,membership),pb_univ) =
                         pullback(1C)(truth,eval)
         val universal =
          fn f =>
           let val ((_,chi,t,_),sc_univ) =
                         character_square(m)
               val f = exp_adjoint(source(C)(chi),chi)
               val fxa = a_prod_o_within(1C)(f,a)
               val u = pb_univ(t,compose(C)(fxa,f))
               val square = (membership,fxa,u,f)
```

```
                    val univ =
                        fn (p,q) => sc_univ(compose(C)(h,p),q)
                 in (f,(square,univ)) end
          in (P_of_a,membership,universal) end end
```

7.2.3   Logic in a topos

In this section we define the internal logic in a topos and program it in
ML. We then run this on an example topos to produce the truth-tables
of a three-valued logic.

Firstly, we need the following result.

Proposition 11 *In a topos, every arrow can be factored as an epi followed by a monic.*

Proof   Given an arrow $f : a \to b$ in topos **T**, let (p, q) be the pushout
of (f, f) and $m : d \to b$ the equalizer of the parallel pair (p, q). The
universality of the equalizer gives a unique arrow $e : a \to d$ such that
$f = m.e$. It can be verified that this is an epi-monic factorization of f.
□

The factorization can be encoded as a function `factorize` taking a
topos and an arrow and returning the two arrows which are its epi-monic
factorization.

Let **T** be a topos with subobject classifier $t : 1 \to \Omega$ and initial
object 0. The internal logic in **T** arises from the following definitions of
the logical connectives:

- *true* $: 1 \to \Omega$ is the arrow t.

- *false* $: 1 \to \Omega$ is the character of the unique arrow $0 \to 1$.

- *not* $: \Omega \to \Omega$ is the character of *false*.

- *and* $: \Omega \times \Omega \to \Omega$ is the character of $< true, true >: 1 \to \Omega \times \Omega$.

- *or* $: \Omega \times \Omega \to \Omega$ is the character of the image (the monic part of
  the epi-monic factorization) of the arrow

  $$[< T, i_\Omega >, < i_\Omega, T >] : \Omega + \Omega \to \Omega \times \Omega$$

  where $T : \Omega \to \Omega$ is constantly true.

- *imply* $: \Omega \times \Omega \to \Omega$ is the character of the equalizer of the arrows
  and, $\pi_1 : \Omega \times \Omega \to \Omega$.

These operations can be directly encoded using limits, colimits and the subobject classifier. Exponentials are used only in defining quantifiers (see Exercise 1):

```
fun FALSE (T as topos(C,lim,colim,_,_)) =
  let val (init,_) = initial(cocomplete_cat(C,colim))
      val f = terminal_arrow(complete_cat(C,lim))(init)
  in character(T)(f) end


fun NOT(T) = character(T)(FALSE T)


fun AND(T as topos(C,lim,colim,_,sc) ) =
  let val lC = complete_cat(C,lim)
      val ((truvals,truth),_) = sc
      val (_,p_univ) = product(lC)(truvals,truvals)
      val m = p_univ(terminal_obj(lC),truth,truth)
  in character(T)(m) end


fun OR(T as topos(C,lim,colim,_,((truvals,truth),_))) =
  let val lC = complete_cat(C,lim)
      val cC = cocomplete_cat(C,colim)
      val ((coprod,p,q),cp_univ) =
              coproduct(cC)(truvals,truvals)
      val ((prod,r,s),pr_univ) =
              product (lC) (truvals,truvals)
      val t =
        compose(C)(truth,terminal_arrow(lC)(truvals))
      val p1 = pr_univ(truvals,identity(C)(truvals),t)
      val p2 = pr_univ(truvals,t,identity(C)(truvals))
      val m  = cp_univ(prod,p1,p2)
      val (_,im_of_m) = factorize(T)(m)
  in character(T)(im_of_m) end

fun IMPLY(T as topos(C,lim,colim,_,((truvals,truth),_)))=
  let val lC = complete_cat(C,lim)
      val ((prod,r,s),pr_univ) =
              product(lC)(truvals,truvals)
      val m = pr_univ(terminal_obj(lC),truth,truth)
      val conj = character(T)(m)
      val ((_,m1),_) = equalizer(lC)(conj,r)
  in character(T)(m1) end
```

7.2.4   An example: a three-valued logic

We could apply these functions to the topos of finite sets to get an extremely contorted method of computing the familiar truth-tables of classical propositional logic. However, the genericity of the code allows more exotic truth-values and internal logics. Let us use this code to generate a three-valued logic.

We define the topos structure of **FinSet**$^{\rightarrow}$, the category whose objects are **FinSet**-arrows and whose arrows are commuting squares. This is the comma category $(I_{\textbf{FinSet}}, I_{\textbf{FinSet}})$.

Alternatively, this category is the functor category from the category with two objects and one non-identity arrow between them. There are general constructions of toposes as functor and comma categories. In the exercises, we suggest that the reader programs some of these constructions. In fact, the functor category description of **FinSet**$^{\rightarrow}$ provides its topos structure through a general construction. However, here we explicitly code this topos, avoiding the overheads arising from representing the general construction. Moreover, we use the comma category representation as it is somewhat more succinct.

Limits and colimits in **FinSet**$^{\rightarrow}$ arise from the description as a comma category:

```
val arrow_cat = comma_cat(I(FinSet),I(FinSet))
val cocomplete_arrow_cat =
      cocomplete_comma_cat
            (cocomplete_FinSet,cocomplete_FinSet)
            (cocontinuous_I(FinSet),I(FinSet))
val complete_arrow_cat =
      complete_comma_cat
            (complete_FinSet,complete_FinSet)
            (I(FinSet),continuous_I(FinSet))
```

The subobject classifier is rather interesting. The object of truth-values is an object in the comma category and so is a pair of sets and a set arrow between them. The source set is a three-element set:

```
val src = {just("F"),just("*"),just("T")}
```

and the target is a two-element set:

```
val tgt = {just("T"),just("F")}
```

The function between them is defined in the following:

```
val Omega =
    ( src,
      set_arrow(src,
                fn just("T") => just("T")
                 | just("*") => just("T")
                 | just("F") => just("F"),
                tgt),
        tgt )
```

It is clear that we get something more than a version of two-valued classical logic. The arrow $t : 1 \to \Omega$ representing 'true' selects the element just("T") in both the source and target sets:

```
val truth =
    let val T_obj as (T,_,T') =
          terminal_obj(complete_arrow_cat) in
    comma_arrow( T_obj,
                (set_arrow(T,fn _ => just("T"),src),
                 set_arrow(T',fn _ => just("T"),tgt)),
                Omega ) end
```

The required property of Ω is that any monic has a characteristic arrow into Ω to make a square into a pullback. Monics in the comma category from (a, f, a') to (c, g, c') are pairs of monics $(m : a \to c, m' : a' \to c')$ such that the following square commutes:

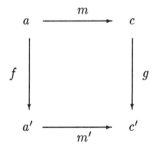

The three-valued source set src is used to define the character of a monic as follows. For an element $z \in c$, if z is in the image of a then it maps to just("T"); otherwise, if it is in the inverse image of a' through g but not in the image of a then it maps to just("*") else it maps to just("F"):

```
fun chi(comma_arrow(s_obj as (a,f,a'),
                    (m,m'),
                    t_obj as (c,g,c'))) =
  comma_arrow(t_obj,
              (set_arrow(c,fn z =>
                  if member(z,image_set m) then just("T")
                  else if member(z,
                    (inv_image(g,image_set(m'))
                        diff image_set(m)))
                  then just("*") else just("F"),src),
                set_arrow(c',fn z =>
                  if member(z,image_set m') then just("T")
                  else just("F"),tgt) ),
              Omega )
```

These components are gathered together, with the universality of the pullback square, in the following definition of the subobject classifier in **FinSet$^{\rightarrow}$**:

```
val subobject_classifier =
    let val 1C as complete_cat(C,_) =
            complete_arrow_cat in
      ( (Omega,truth),
        fn M as comma_arrow(s_obj as (a,f,a'),
                            (m,m'),
                            t_obj as (c,g,c')) =>
            let val t =
                terminal_arrow(1C)(source(C)(M)) in
            ((truth,chi(M),t,M),
             fn (p,q) =>
                compose(C)
                (comma_arrow(target(C)(q),
                            (inv(m),inv(m')),
                            s_obj),q) )
            end   ) end
```

This is all that is required to define the truth-tables of the logical connectives (exponentials appear only when defining quantifiers). Let us call the topos arrow_topos.

We can now run the programs above to compute the logical connectives. Arrows $1 \rightarrow \Omega$ are in 1-1 correspondence with elements of the set src and, moreover, functions on Ω are characterized by the action on these elements. We thus consider this to be a three-valued logic whose truth-values are the elements of src and we display the resulting connectives as truth-tables.

First, the value of FALSE(arrow_topos) is the arrow from the terminal object selecting the element just("F") in each set of Ω.

The value of NOT(arrow_topos) is an arrow in the comma category from Ω to Ω. Its action on the set src is given in the following table:

NOT	
T	F
*	F
F	T

The connectives AND(arrow_topos) and OR(arrow_topos) are both arrows in the comma category from $\Omega \times \Omega$ to Ω. The results of calculating these values are given in the following tables:

AND	T	*	F
T	T	*	F
*	*	*	F
F	F	F	F

OR	T	*	F
T	T	T	T
*	T	*	*
F	T	*	F

Finally, here is the truth-table of implication, calculated as the value of IMPLY(arrow_topos):

IMPLY	T	*	F
T	T	*	F
*	T	T	F
F	T	T	T

This is a simple temporal logic discussed in [Lawvere 76].

Let us examine the ingredients in these calculations. Explicitly, we use products and coproducts in the comma category, and implicitly, in the factorization, we use equalizers and pushouts. These are computed from those in **FinSet** using the inheritance of colimits to comma categories and, through the encoding of duality, the inheritance of limits. As well as this, the image and inverse image operations on set arrows are called from the subobject classifier. These also are computed using limits and colimits in **FinSet**. All this to calculate the simple truth-tables above!

7.3   Conclusion

So far we have displayed a good deal of category theory in the programming language ML and shown explicitly how categorical constructions yield programs. We have travelled fairly methodically through standard constructions and run the resulting programs on examples which we hope have been of some interest to the reader. Where do we go from here?

Because of the generality of category theory and its applicability in set theory, algebra, topology and logic, it may be expected that the categorical programs we have developed would similarly be of widespread application. However, it turns out that finding applications is not as easy as we at first envisaged. Programming tasks are often described at a fairly low level of abstraction and in order to use the programs we have developed we must interpret the tasks at a much more abstract level – as instances of categorical concepts. Moreover, we then must find relevant constructions in category theory to provide algorithms. The following two chapters are case studies in the application of categorical programming, first to developing unification algorithms and then to implementing a semantics.

There are certain categorical constructs which we have not encoded as they cannot be represented within the type system of ML at a sufficient level of generality. An example of such a construct is indexed categories which occur, for instance, as hyperdoctrines in categorical logic. Another example is the general term algebra construction which, we have presented only for finite algebras. Both of these require explicit type universes. We are now engaged in the development of a suitable type system for categorical programming. We discuss linguistic aspects of category theory at length in Chapter 10.

7.4 Exercises

Exercise 1. Quantifiers Universal and existential quantifiers can be defined in a topos. Using definitions from standard texts, e.g. [Goldblatt 79], verify that the following ML functions define the quantifiers:

```
fun FOR_ALL(T)(a) =
    character(T)(name(T)(true_(T)(a)))
fun EXISTS (T as topos(C,lim,_,_,_))(a) =
    let val (P_of_a,a_membership,_) = power(T)(a)
        val ((_,pr,_),_) =
          product(complete_cat(C,lim))(P_of_a,a)
        val f = compose(C)(pr,a_membership)
        val (_,im_of_f) = factorize(T)(f)
    in character(T)(im_of_f) end
```

Exercise* 2. Constructing toposes: 1. Slice categories The following is called the fundamental theorem of toposes: If **C** is a topos, then so is the slice category **C**/*a* for any object *a* of **C**.

Recall that the slice category **C**/*a* has as objects, arrows into *a* and as arrows, commuting triangles. Details of the topos structure of slice categories are found in [Freyd 72] and [Goldblatt 79]. You may like to program this as a way of generating some interesting toposes.

Exercise* 3. Constructing toposes: 2. Functor categories For any small category **B** (i.e. category whose collections of objects and arrows are both sets), the functor categories **Set**$^{\mathbf{B}}$ and **FinSet**$^{\mathbf{B}}$ are toposes. The objects in these categories are called presheaves.

The details of this construction may be found in [Freyd 72], [Goldblatt 79] and [Barr, Wells 85]. The subobject classifier is constructed as sets of 'sieves'. Again the construction can be expressed as a program for generating toposes.

Exercise* 4. A topos of graphs Let **B** be the category with two objects *s* and *t* and two arrows both from *s* to *t* as well as the identity arrows. Then the functor topos **FinSet**$^{\mathbf{B}}$ is the topos of finite graphs.

Investigate the topos structure of graphs in this topos. *Hint*: the subobject classifier has two nodes *a* and *b* and five edges. Two edges from a to itself and one from b to itself; the other two are between a and b in opposite directions.

Chapter 8

A Categorical Unification Algorithm

We turn from programming basic category theory to applications. In this chapter, we derive an algorithm for the unification of terms from constructions of colimits. This is a case study in using the abstraction and constructivity of category theory in the design of algorithms[1].

Unification is a symbol-manipulative task occurring in many areas of computation, in particular in the automation of inference. For those not familiar with unification we describe it below. Unification is an equation-solving task that is decidable. The earliest algorithm for computing most general unifiers was given by Herbrand [1930]. Robinson [1965] applied this to automated inference. Since then many variations on the basic unification algorithm have been proposed, often to improve efficiency. The algorithm we present is a non-deterministic version of the basic algorithm of Robinson.

The derivation of a unification algorithm hinges upon two observations. Firstly, unification can be considered as an instance of something more abstract – as a colimit in a suitable category. Secondly, general constructions of colimits provide recursive procedures for computing the unification of terms.

As it stands, this derivation of a unification algorithm is an isolated result. However, it finds a setting in a categorical treatment of equational deduction. This is described in [Rydeheard, Stell 87] where we draw upon categorical logic to give a 2-category structure for equational deduction. In this setting, we consider solving equations in equational theories and combining unification algorithms for different theories.

8.1   The unification of terms

As an example of unification, consider the following equation between terms:

$$f(w, g(h(y)), h(z)) = f(g(x), z, h(w))$$

Here f, g and h are operator symbols with the evident arities and w, x, y, z are variables. The task of unification is to replace the variables with terms so that both sides of the equation become the same term. Such a substitution is called a *unifier*. For instance, the substitution,

$$w \mapsto g(h(v)), \quad x \mapsto h(v), \quad y \mapsto v, \quad z \mapsto g(h(v))$$

where v is a variable, makes both sides of the equation equal to

$$f(g(h(v)), g(h(v)), h(g(h(v))))$$

Not only is this a unifier, it is the 'most general unifier' in that any other unifier factors through it.

Unifiers need not always exist for an equation. We can distinguish two cases when they fail to exist. As examples:

- *Clash* $g(x) = h(y)$

- *Cyclic* $x = g(x)$

In the 'clash' case no substitution can possibly make the two sides equal. However, in the 'cyclic' case unifiers do exist if we allow infinite terms.

In programming, unification occurs, for instance, in computational logic [Robinson 65], polymorphic type-checking [Milner 78] and in implementing programming languages which are based upon pattern-matching such as Prolog [Colmerauer et al. 73]. A good general survey of term rewriting and unification is [Huet, Oppen 80]. Efficient unification algorithms have been proposed, for instance those of [Paterson, Wegman 78] and of [Martelli, Montanari 82]. Unification admits several generalizations including higher-order unification [Huet 75] and unification in equational theories [Huet, Oppen 80], [Siekman 84].

---

[1]This account appears in the *Proc. Summer Conf. on Category Theory and Computer Programming (1985)*, *LNCS* 240, Eds. D. Pitt et al., Springer-Verlag 1986. A full account of the material of this chapter may be found in internal reports of the Departments of Computer Science in the Universities of Manchester and Edinburgh.

Manna and Waldinger [1980], Eriksson [1984] and Paulson [1985] have considered derivations of unification algorithms and the reader is invited to compare this categorical version with theirs. The algorithm we are to derive is a general recursive counterpart of the non-deterministic algorithm in [Martelli, Montanari 82] which there serves as a starting point for the development of an efficient algorithm. It is an open question as to whether such efficient evaluation strategies can be understood in this categorical framework. It is the compositional structure of unification algorithms rather than their efficiency which seems open to this abstract analysis.

8.2   Unification as a coequalizer

We recall the material of Chapter 3 on terms and term substitutions. Let Ω be an operator domain. Denote by $T_\Omega(X)$ the set of terms over Ω in variables from the set X. A (term) substitution $f : X \to Y$ is a function

$$f : X \to T_\Omega(Y)$$

Substitutions may be applied to terms, replacing variables by terms. They form arrows in a category $\mathbf{T}_\Omega$ whose objects are sets. The full subcategory of finite sets we denote by $\mathbf{T}_\Omega{}^{Fin}$.

We show that the task of unification is exactly that of computing coequalizers in the category $\mathbf{T}_\Omega$ (an observation made by Goguen).

An *equation* in set X over Ω is a pair of terms (s,t), which we write as $s = t$. A substitution $q : X \to Y$ is said to *unify* a set of equations in X, $\{s_i = t_i : i \in I\}$, if $\forall i \in I$. $q(s_i) = q(t_i)$. Such unifiers do not always exist. However, when they do exist so does a *most general unifier* defined to be a unifier $q : X \to Y$ such that for any unifier $q' : X \to Y'$ there is a unique[2] substitution $u : Y \to Y'$ satisfying $uq = q'$.

Now, a set of equations in X over Ω, $\{s_i = t_i : i \in I\}$, is equivalent to a parallel pair of arrows in $\mathbf{T}_\Omega$,

$$I \underset{g}{\overset{f}{\rightrightarrows}} X$$

defined by $f(i) = s_i$ and $g(i) = t_i$. Unifiers of this set of equations are arrows q such that $qf = qg$. Moreover, the above definition of a most general unifier is exactly that of a coequalizer of f and g in $\mathbf{T}_\Omega$.

---

[2]Uniqueness is often not demanded here but, since epis are involved, uniqueness is assured.

Notice how, in the above, the universal form of the definition of the most general unifier translates directly into the universal definition of the coequalizer.

An alternative, somewhat more standard way of treating the compositional structure of unification goes as follows. Fix an infinite (global) set of variables X. Instead of considering substitutions to be maps between sets, let them be maps between terms in X. We then form a category whose objects are terms and whose arrows are substitutions. This category is sometimes reduced to a pre-order ($s \lesssim t$ iff t is a substitution instance of s) and, from this, a lattice, the lattice of subsumptions [Plotkin 69], may be constructed. The most general unifier of two terms is their least upper bound in this pre-order.

The localization of variables in our treatment seems crucial. It allows us to use general constructions *at the level of category theory* as a source of unification algorithms. Moreover, it avoids much of the explicit handling of variables which occurs in standard treatments of term-rewriting – indeed, even when using a global set of variables, there are occasions, such as in defining 'narrowing', when local sets of variables are required. Localization affords a uniform treatment of variables, as they are handled automatically within the limit and colimit operations in $\mathbf{T}_\Omega$.

8.3   On constructing coequalizers

As we have remarked previously, category theory is particularly rich in ways of constructing colimits from other colimits. For unification algorithms we need constructions of coequalizers. The following two theorems provide constructions, in an arbitrary category, of coequalizers in terms of other coequalizers. Amongst the many possible such theorems, these two are chosen so as to lead to a recursive algorithm which terminates in the category $\mathbf{T}_\Omega{}^{Fin}$ and hence provides unification algorithms. Of course, neither is particularly 'deep' nor are they new but together they reflect the compositional structure of unification.

Each theorem can readily be verified by simple arrow-chasing. In fact, the second theorem follows directly from the definition of an epi.

The first theorem considers parallel pairs of arrows whose source can be expressed as a coproduct, whilst the second theorem deals with parallel pairs of arrows that can be factored through a common arrow. In the case of unification (i.e. in the category $\mathbf{T}_\Omega$) the first theorem corresponds to the division of the set of equations into two parts (we have met it already in Chapter 4), whilst the second theorem corresponds to the division of terms into subterms.

Theorem 9 *If $q : b \to c$ is a coequalizer of the parallel pair,*

$$a \overset{f}{\underset{g}{\rightrightarrows}} b$$

and $r : c \to d$ is the coequalizer of

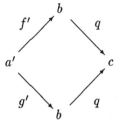

then $rq : b \to d$ is the coequalizer of the following parallel pair:

$$a + a' \overset{[f,f']}{\underset{[g,g']}{\rightrightarrows}} b$$

Theorem 10 *For all epis $h : a' \to a$, the arrow $q : b \to c$ is a coequalizer of the parallel pair of arrows $f, g : a \to b$ iff it is a coequalizer of the parallel pair:*

$$a' \overset{fh}{\underset{gh}{\rightrightarrows}} b$$

It is to be emphasized that these theorems are valid for any category. However, for our present purposes, we illustrate them in the category $\mathbf{T}_\Omega$. Consider the following two equations (with f, g, h, a operators and w, x, y, z variables):

$$f(w, g(h(y)), h(z)) = f(g(x), z, h(w))$$

$$h(a) = x$$

The most general unifier q of the first equation is given previously as

$$w \mapsto g(h(v)), \; x \mapsto h(v), \; y \mapsto v, \; z \mapsto g(h(v))$$

The second equation with q applied to it is

$$h(a) = h(v)$$

Its most general unifier r is simply

$$v \mapsto a$$

According to Theorem 9, the most general unifier of the two equations is then rq, which is the following substitution:

$$w \mapsto g(h(a)), \ x \mapsto h(a), \ y \mapsto a, \ z \mapsto g(h(a))$$

Theorem 10 says that, for instance, the most general unifier of the two equations above is the same as the most general unifier of the following set of equations obtained by matching subterms:

$$w = g(x)$$

$$g(h(y)) = z$$

$$h(z) = h(w)$$

$$h(a) = x$$

There is a calculus of parallel pairs based on the following two operations. Letting $+$ be a distinguished coproduct, define the *coalesced sum* of parallel pairs as

$$\left(a \overset{f}{\underset{g}{\rightrightarrows}} b \right) \oplus \left(a' \overset{f'}{\underset{g'}{\rightrightarrows}} b \right) \ = \ a + a' \overset{[f,f']}{\underset{[g,g']}{\rightrightarrows}} b$$

Define the *right composition* of an arrow with a parallel pair as

$$\left(a \overset{f}{\underset{g}{\rightrightarrows}} b \right) \ \circ \ \left(a' \overset{h}{\longrightarrow} a \right) = \ a' \overset{fh}{\underset{gh}{\rightrightarrows}} b$$

The *left composition* (also denoted $\circ$) is defined similarly.

In terms of these operations, the coequalizer function ϕ taking parallel pairs to arrows satisfies the following equations (in the sense that if the right side is defined so is the left and they are equal):

$$\phi(P \oplus Q) = \phi(\phi(P) \circ Q)\phi(P)$$

$$\phi(P \circ h) = \phi(P) \quad (h \text{ an epi})$$

These equations are simply a rewriting of Theorems 9 and 10. Notice that they do not express ϕ as a homomorphism.

Let us say that a parallel pair is *irreducible* if it cannot be expressed non-trivially as a coalesced sum or as a right composite. An expression for a parallel pair P as $Q \oplus R$ is trivial if either Q or R is isomorphic to P. Likewise, an expression for P as $Q \circ h$ is trivial if Q is isomorphic to P.

The following theorem says that in the category $\mathbf{T}_\Omega{}^{Fin}$ these equations define a function – the coequalizer.

Theorem 11 *Let $\mathcal{C}$ be the class of coequalizable parallel pairs in $\mathbf{T}_\Omega{}^{Fin}$. There is a unique (to within an isomorphism) function,*

$$\phi : \mathcal{C} \to Arrow(\mathbf{T}_\Omega{}^{Fin})$$

defined to be the coequalizer on irreducible parallel pairs and satisfying the following equations (in the sense that if the right side is defined so is the left and they are equal):

$$\phi(P \oplus Q) = \phi(\phi(P) \circ Q)\phi(P)$$

$$\phi(P \circ h) = \phi(P) \quad (h \text{ an epi})$$

Moreover, $\phi(P)$ is the coequalizer of $P \in \mathcal{C}$.

Sketch of proof

The universality of the arrow $\phi(P)$, if it exists, is a direct consequence of Theorems 9 and 10 above and hence is at the level of general categorical skull-duggery. To establish the existence of $\phi(P)$ when P is coequalizable is more intricate – unduly intricate compared with the elegance of Theorems 9 and 10. Other authors, e.g. [Manna, Waldinger 80], have noticed this disparity between derivation and proof of unification algorithms.

The existence of $\phi(P)$ is a termination proof and depends on defining a suitable well-founded pre-order. We give a rough argument for termination, ignoring the partial nature of unification.

Define a well-founded pre-order on sets of equations as the lexical product[3] of the pre-orders (i) the number of variables in the set of equations, (ii) the number of occurrences of operators and (iii) the number of equations.

---

[3]The lexical product $\lesssim$ of pre-orders $\lesssim_1$ and $\lesssim_2$ is defined by $x < y$ iff $x <_1 y$ or ($x \sim_1 y$ and $x <_2 y$), together with $x \sim y$ iff $x \sim_1 y \sim_2 x$. The lexical product of well-founded pre-orders is well-founded. The pre-order determined by numerical functions is that corresponding to the usual numerical order.

Now let E be a set of equations. Consider the construction of Theorem 9. Divide E non-trivially into the union of E_1 and E_2. The set E_1 is smaller than E in the pre-order since E_1 has no more variables or operator occurrences than E and E_1 has strictly fewer equations (by non-triviality). Let q be the most general unifying substitution of E_1 and let $q \circ E_2$ be the set of equations resulting from applying q throughout E_2. There are two cases. If q is an isomorphism, then $q \circ E_2 \equiv E_2$ and so, as before, $q \circ E_2$ is smaller than E. If q is not an isomorphism then it reduces the number of variables (an observation of Robinson [1965]) and so again $q \circ E_2$ is smaller than E.

Consider now the construction of Theorem 10. In this case E is expressed non-trivially as $E' \circ h$ with h an epi. E' is smaller than E in the pre-order since both have the same number of variables but the number of operator occurrences in E' is strictly smaller than that in E or is the same but then E' contains fewer equations than E.

This rough argument can be cast into a categorical proof by axiomatizing suitable properties of the category $\mathbf{T}_\Omega^{Fin}$, principally the support of an appropriate well-founded pre-order. It has been pointed out to us that there is a possibility of an entirely categorical termination proof using a theory of ordinals in categorical logic.

8.4   A categorical program

In this section we express the above two constructions of coequalizers as programs. These programs are then used to define a recursive unification algorithm.

In Chapter 4 we deal with computational representations of various colimits including coequalizers. We simplify that treatment here by omitting the universal part of the coequalizer (it can easily be added and will then provide the universal part of the most general unifier). Thus we make the coequalizer a function of type:

```
coequalize: 'a * 'a  -> 'a
```

Here 'a is the type variable representing the type of arrows in the category of interest.

Theorem 9 as a construction, takes two parallel pairs $f, g : a \to b$ and $f', g' : a' \to b$ (using the notation of the theorem) and yields an arrow –

a coequalizing arrow of the parallel pair

$$a + a' \overset{[f,f']}{\underset{[g,g']}{\rightrightarrows}} b$$

The type of this function is:

```
sum_coequalize: ('o,'a)Cat -> (('a * 'a -> 'a) ->
                (('a * 'a) * ('a * 'a) -> 'a))
```

Referring back to the statement of Theorem 9, we interpret this as: in a category with coequalizers, parallel pairs (f, g) and (f', g') determine an arrow (rq in the notation of the theorem). The definition of the function is simply a rewording of the theorem:

```
fun sum_coequalize(C)(cq)((f,g),(f',g')) =
  let val q = cq(f,g)
      val r = cq(compose(C)(q,f'),compose(C)(q,g')) in
  compose(C)(r,q) end
```

Theorem 10 is dealt with similarly, giving the following trivial program which says simply that the coequalizer of a parallel pair $P \circ h$ (with h an epi) is that of the parallel pair P.

```
fun composite_coequalize(C)(cq)(P,h) = cq(P)
```

It is Theorem 11 and its proof that is to provide unification algorithms. Looking back at the theorem, we see that, apart from the equations defining coequalizers of reducible parallel pairs and the direct definition of coequalizers of irreducible parallel pairs, there remains only the constructive interpretation of reducibility, i.e. encoding the following:

$$\exists Q, R : \ P \equiv Q \oplus R, \ \ Q \not\equiv P \not\equiv R$$

$$\exists Q, h \ (epi) : \ P \equiv Q \circ h, \ \ h \ not \ an \ isomorphism$$

This becomes a pair of decomposition functions which, if supplied with reducible parallel pairs, give the components as results. On other arguments these functions may not be defined – in programming terms they may 'fail' or 'raise an exception'. We should consider here some form of non-determinism as, for the correctness of the algorithm, it is immaterial in what order the decomposition takes place. Moreover, the same applies to the order in which the equations (and hence the decompositions) are chosen. Clearly, some strategies are more efficient than others and there remains the question of how to constrain the algorithm to efficient strategies.

Recall from Chapter 3 the computational representation of the category $\mathbf{T}_\Omega{}^{Fin}$. There we introduced terms and defined term substitutions as:

```
datatype Substitution =
  subst of (element Set)*(element -> Term)*(element Set)
```

By introducing functions for the composition and identity of such substitutions we defined a category `FinKleisli`.

We are now ready to describe the decomposition of parallel pairs of substitutions. To express a parallel pair non-trivially as a coalesced sum we express the source set as the disjoint union of non-empty sets, i.e. non-trivially as a coproduct in $\mathbf{T}_\Omega{}^{Fin}$. If this is not possible we raise an exception. The appropriate function is defined below using a function `split` which expresses a finite set (cardinality greater than one) as a non-trivial coproduct in $\mathbf{T}_\Omega{}^{Fin}$ (this can be done in various ways).

```
fun sum_decompose(P) =
    if cardinal(pp_source P) > 1 then
        let val (b,c) = split(pp_source P) in
        (restrict(b,P),restrict(c,P)) end
    else raise sum_decompose
```

The function `restrict` takes a subset a' of the source of a parallel pair P of substitutions f, g and yields the parallel pair of f and g restricted to a'.

Expression as a non-trivial right composition is more complex, involving the factorization of terms. One way of factoring a term $\rho(t_1, t_2)$ in X is to express it as a term $\rho(z_1, z_2)$ in $Z = \{z_1, z_2\}$, the arity of ρ, together with a substitution $p : Z \to X$ defined by $p(z_1) = t_1$ and $p(z_2) = t_2$. Note that the substitution $s = \lambda x.\rho(z_1, z_2) : 1 \to Z$ is an epi.

This extends to a factorization of substitutions as follows. A (non-variable) term in X is a pair $\langle \rho, p : Z \to X \rangle$ with Z the arity of ρ and p a substitution. A substitution contains a term $\langle \rho, p : Z \to X \rangle$ iff it is (to within an isomorphism) of the form $[f', ps] : Y' + 1 \to X$ where $s : 1 \to X$ is $s = \lambda x.\rho(i_X)$. This substitution may then be factored in $\mathbf{T}_\Omega$ as:

$$Y' + 1 \xrightarrow{i_{Y'} + s} Y' + Z \xrightarrow{[f', p]} X$$

Note that $i_{Y'} + s$ is an epi. Moreover, as long as we factor only pairs of terms with a common leading operator, this factorization extends to the right factorization of parallel pairs of substitutions.

Coproducts in the category $\mathbf{T}_\Omega^{Fin}$ are disjoint unions and can be programmed as in **FinSet** using a labelling with, say, 'pink' and 'blue' to ensure disjointness.

We shall encode this decomposition as a program – firstly defining a function which checks whether a pair of terms has the same leading operator by case analysis on the form of the terms:

```
fun top_same(apply(phi,_),apply(psi,_)) = phi=psi
 | top_same(s',var(x)) = false
 | top_same(var(y),t') = false
```

The next function identifies such a pair of terms in a parallel pair of substitutions:

```
fun witness(P as (f,g)) =
      if is_empty(pp_source P) then raise witness else
      let val (x,a') = singleton_split(pp_source(P)) in
   if top_same(subst_apply(f)(x),subst_apply(g)(x))
      then x else witness(restrict(a',P)) end
```

Here `singleton_split` extracts an element from a set and returns the element and the remaining set. We can now define the decomposition function which expresses a parallel pair of substitutions, if possible, as a right composite $Q \circ h$ returning the epi h and the parallel pair Q as results.

```
fun composite_decompose(P as (f,g)) =
  let val x = witness(P)
      val apply(phi,s) = subst_apply(f)(x)
      and apply(psi,t) = subst_apply(g)(x)
      val c = sum(minus(pp_source(P),x),arity(phi)) in
  ((subst(c,
          fn pink(z) => subst_apply(f)(z)
          | blue(z)=> s(z),
          pp_target(P)),
    subst(c,
          fn pink(z) => subst_apply(g)(z)
          | blue(z)=> t(z),
          pp_target(P)) ),
```

```
subst(pp_source(P),
      fn z => if x=z
          then apply(phi,fun y => var(blue(y)))
          else var(pink(z)),
      c)) end
```

Here pp_source and pp_target are the source and target functions for parallel pairs of substitutions, minus subtracts an element from a set and subst_apply applies a substitution to an element of its source yielding a term.

The final task before giving a recursive unification algorithm is the computation of coequalizers of irreducible parallel pairs of substitutions. Irreducible parallel pairs which have coequalizers are either empty, in which case the coequalizer is an identity, or are equivalent to a pair of terms $(\langle x \rangle, t)$ in X with $x \in X$ and $x \notin Var(t)$ (unless t is a variable). The coequalizer then is simply the substitution which takes x to t and acts as identity on all other elements of X. The first function below creates this substitution. The second function unifies irreducible parallel pairs when possible.

```
fun unit_unify(x,t,b,c) =
      subst(b,fn z => if z=x then t else var(z),c)

fun irreducible_unify(P as (f,g)) =
          let val b = pp_target P in
      if is_empty(pp_source P)
          then subst_identity(b)
          else
          let val (x,_) = singleton_split(pp_source P)
              val s = subst_apply(f)(x)
              and t = subst_apply(g)(x) in
          case (s,t) of
            (var(z),var(z')) =>
                let val c =
                    if z=z' then b else minus(b,z) in
                unit_unify(z,t,b,c) end |
```

```
(var(z),t') =>
  if occurs(z,t') then raise cyclic else
    unit_unify(z,t',b,minus(b,z)) |
(s',var(z)) =>
  if occurs(z,s') then raise cyclic else
    unit_unify(z,s',b,minus(b,z)) |
(_,_)        => raise clash end end
```

The function occurs checks whether a variable occurs in a term.

Finally, a program for the most general unifier of a parallel pair. This is a case analysis – express the pair, if possible, non-trivially as a coalesced sum or as a right composite and calculate the unifier by Theorem 9 or Theorem 10 calling unification recursively. If decomposition is not possible then the function above calculates unifiers of irreducible parallel pairs. Exception handling is used to check the form of a parallel pair by attempting the decomposition. If the decomposition fails then a handler is invoked which either attempts another decomposition or deals with the irreducible cases.

```
fun unify(P) =
  let val (P,P') = sum_decompose(P) in
      sum_coequalize(FinKleisli)(unify)(P,P') end
    handle ? =>
  let val (P',h) = composite_decompose(P) in
      composite_coequalize(FinKleisli)(unify)(P',h) end
    handle ? =>
      irreducible_unify(P)
```

This then is a program for unification based on constructions of colimits. It could be expressed using conditionals rather than exceptions but the tests for decomposability proceed by attempting the decomposition, so code would be repeated in the test and the decomposition. The universal property of the most general unifier could be included by encoding it for the two functions derived from Theorems 9 and 10, and amending the code for irreducible parallel pairs.

Exercise* 1. Consider a parallel pair of arrows $f, f' : a \rightarrow b$. Suppose $f = hg$. Then we may take the pushout p, p' of g, f' and the coequalizer q of $p, p'h$. Show that the coequalizer of f, f' is qp'.

In a category with coproducts, this provides a recursive construction of coequalizers based on the factorization of one arrow in a parallel pair rather than the simultaneous factorization of both. Does this provide a unification algorithm i.e. does the recursion terminate in $\mathbf{T}_\Omega{}^{Fin}$?

Exercise* 2. Inductive generalization There is a dual to unification called generalization by Plotkin [1969]. A *common instance* of two terms s and t is a term u such that there are substitutions f and g with $f(u) = s$ and $g(u) = t$. A *generalization* of two terms is a greatest common instance. This is dual to unification in the lattice of subsumptions. Does generalization have a categorical formulation in $\mathbf{T}_\Omega$ and what about algorithms for generalization?

Chapter 9

Constructing Theories

Computer scientists have been concerned over the last ten years or so with the question of how to specify a task before embarking on the design of a program to perform that task. Informal specifications in English are vague, so a lot of research has gone into the design of formal specification languages. For example, if we want to write a matrix inversion program we might formally specify the reals and operations on them, matrices, the identity matrix and matrix multiplication; we would then be able to say what it means to compute the inverse of a matrix. If we want to write a compiler which translates Pascal into assembly code we need precise definitions of both these languages. Specification languages may be based on equations, on predicate calculus or on a higher order logic. A specification consists of a sequence of declarations of types (or sorts) and operations over them, together with a set of equations or logic sentences. The closure of this set under the appropriate inference operations forms a theory. Thus the specification is what is called in algebra a 'presentation' of the theory.

A long, unstructured list of axioms (equations or logic sentences) is prone to error and hard to understand. So a number of proposals have been made for modular specification languages in which a structured specification may be given using operations to combine smaller pieces of specification. This is analogous to the structuring operations in programming languages such as Modula2 or Ada. We want to define specifications with parameters which can later be filled in. For example, the notion of matrix and the operations over matrices can be generalized to work over any ring, just as the notion of sorting a sequence can be defined for any ordering. These combination operations are really independent of the particular language used to state the axioms, and they can be defined for any language which permits some substitution operations on sentences which are 'semantically well behaved' in some elementary sense. To define the operations independent of the base language, Goguen and Burstall [1983] formulated the concept of 'institution' to explicate the above criterion. It turned out to be very close to logicians' work on abstract model theory, initiated by Barwise [1974].

Burstall and Goguen [1980,81] describe a specification language, called Clear. Its semantics is given in terms of institutions. The operations for building specifications are interpreted as colimits in a suitable category. It this chapter, we show how to implement these semantic operations using the preceding categorical programming. We begin by describing the specification-building operations at a syntactic level. We then define institutions and the semantics of the operations. We then code up the appropriate colimits and implement the semantics.

9.1   Preliminaries

As a preliminary to looking at the construction of algebraic specifications, we review some fairly standard universal algebra, extending that of Section 3.2.7 to the many-sorted (or 'heterogeneous') case.

For S a set, an S-sorted set X is a family of sets X_s, $s \in S$. A *signature* is a many-sorted operator domain. If S is a set of sorts, then a signature is a pair $\Sigma = (S, (O, a))$ where S is a set (of sorts), O is a set (of operators) and a assigns to each operator a non-empty sequence of sorts in S. For $\rho \in O$ with $a(\rho) = s_1, \ldots, s_n, s$, we write $\rho : s_1, \ldots, s_n \to s$.

As an example of a many-sorted signature, consider that of graphs:

```
Graph_Sig =
    sorts node, edge
    opns  source : edge -> node
          target : edge -> node
```

and also that of edge-labelled graphs:

```
Labelled_Graph_Sig =
    sorts node, edge, element
    opns  source : edge -> node
          target : edge -> node
          label  : edge -> element
```

An arrow from signature $\Sigma = (S, (O, a))$ to signature $\Sigma' = (S', (O', a'))$ is a pair of functions $f : S \to S'$, $g : O \to O'$ such that, for each oper-

ator $\rho \in O$, if $a(\rho) = s_1, \ldots, s_n, s$ then $a'(g(\rho)) = f(s_1), \ldots, f(s_n), f(s)$.
Signature arrows are composed componentwise to form a category **Sign**.

An example of a signature arrow,

$$\texttt{s : Labelled\_Graph\_Sig -> Graph\_Sig,}$$

maps the sort **element** to the sort **node**, the operator **label** to **source** and is identity elsewhere.

For X an S-sorted set, we define the S-sorted set of terms $T_\Sigma(X)$ as:

$$x \in X_s \Rightarrow \langle x \rangle \in T_\Sigma(X)_s$$

$$\rho : s_1, \ldots, s_n \to s \ \text{ and } \ t_i \in T_\Omega(X)_{s_i}, 1 \le i \le n \Rightarrow$$

$$\rho(t_1, t_2, \ldots, t_n) \in T_\Sigma(X)_s$$

An equation in variables X is a pair of terms in $T_\Sigma(X)$ with the same sort. A signature arrow may be applied to a term, and hence to an equation, by translating the operators in the term through the signature arrow.

An equational presentation of a theory is a pair $(\Sigma, \mathcal{E})$ of a signature Σ and a set $\mathcal{E}$ of Σ-equations.

An example of an equational presentation is that of graphs with distinguished loops at each node:

```
Loop_Graph =
    sorts node, edge
    opns  source : edge -> node
          target : edge -> node
          loop   : node -> edge
    eqns  source(loop(n)) = n
          target(loop(n)) = n
```

An algebra of a signature $\Sigma = (S, (O, a))$ is an S-sorted set A and for each operator in O, $\rho : s_1, \ldots, s_n \to s$, a function $f_\rho : A_{s_1} \times \ldots \times A_{s_n} \to A_s$. Homomorphisms are a many-sorted version of the single-sorted case, as is the concept of an algebra satisfying an equation. We therefore define categories **Alg**$_\Sigma$ of algebras of a signature Σ, and **Alg**$_{(\Sigma, \mathcal{E})}$ of Σ-algebras satisfying set of equations $\mathcal{E}$. Both the categories **Alg**$_\Sigma$ and **Alg**$_{(\Sigma, \mathcal{E})}$ have initial and free algebras constructed as sorted versions of term algebras.

As examples, algebras of **Graph_Sig** and of **Labelled_Graph_Sig** are graphs and edge-labelled graphs respectively.

A set of Σ-equations $\mathcal{E}$ can be closed with respect to validity, forming a set of equations $\mathcal{E}^*$ with $e \in \mathcal{E}^*$ iff e is satisfied by every algebra in $\mathbf{Alg}_{(\Sigma,\mathcal{E})}$. This coincides with closure under many-sorted equational deduction [Goguen, Meseguer 85], so that $e \in \mathcal{E}^*$ iff e is deducible from the equations $\mathcal{E}$. A set of equations $\mathcal{E}$ is *closed* if $\mathcal{E} = \mathcal{E}^*$. A *theory* is a presentation in which the set of equations is closed. A presentation $(\Sigma, \mathcal{E})$ presents the theory $(\Sigma, \mathcal{E}^*)$. A theory arrow from $(\Sigma, \mathcal{E})$ to $(\Sigma', \mathcal{E}')$ is a signature arrow $F : \Sigma \to \Sigma'$ such that, if $e \in \mathcal{E}$, then $F(e) \in \mathcal{E}'$. We thus form the category of theories $\mathbf{Th}$.

This summarizes the universal algebra that we need to describe the semantics of combining theories. More details of many-sorted algebra may be found in, for instance, [Goguen, Burstall 84].

9.2   Constructing theories

We describe some operations for constructing theories, which we illustrate with equational theories though, as we shall see later, they may be extended to other kinds of theories. These operations were introduced by Burstall and Goguen [1977,80,81] in the specification language Clear.

Combine

The combine of two theories T_1, T_2, written $T_1 + T_2$, is their disjoint union. More properly, it is the theory presented by the disjoint union of the signatures together with the two sets of equations translated into the disjoint union signature. This turns out to be a coproduct in the category of theories.

As an example, consider the two theories presented as follows:

```
const SemiGroup =
    theory sorts element
            opns  _ * _ : element,element -> element
            eqns  (a*b)*c = a*(b*c)
    endth

const Graph =
    theory sorts node, edge
            opns  source : edge -> node
                  target : edge -> node
    endth
```

Then the theory SemiGroup + Graph contains the three sorts and the three operations above satisfying the one equation. More generally, renaming would take place to separate sorts and operations which share the same name.

We are concerned not only with theories but also with specifications – descriptions of the construction of theories from component theories. Consider two specifications of theories T_1, T_2 which contain some common component theory T. Does $T_1 + T_2$ contain one or two copies of T? In practice, it seems more useful to allow only one copy of T (the alternative is called 'proliferation'). This is handled by introducing the concept of 'theories built from theories', which are described as cocones in the category of theories. The combine operation is then interpreted as a coproduct of cocones.

Enrich

Enriching a theory is simply the addition of new sorts, operations and equations and is interpreted as a theory arrow from the original theory to the enriched theory.

As an example, suppose there is a theory of truth-values called Bool (defined later) containing a sort of truth-values, bool, and the constant operations true and false as well as conjunction and implication. Then a theory of ordered sets (in this case, the order is a pre-order) can be described as an enrichment of Bool.

```
const Order =
    Bool enriched by
        sorts element
        opns  _ =< _ : element,element -> bool
        eqns x =< x = true
             x =< y and y =< z implies x =< z = true
    enden
```

Notice that the added equations may refer to operations in the theory being enriched (Bool in this case).

Parameters

Consider graphs whose edges are labelled. For many applications, e.g. for 'flow' problems, it does not matter what the labels are. They might be integers, reals, pairs of integers or even words, as long as the labels support some kind of order. We thus want to abstract the notion of labelled graphs from particular kinds of labels. To do this we treat the labels as a parameter and ensure that they support a requisite order.

Consider graphs with maximum flow values on the edges and an actual flow along the edges. The theory of such graphs may be written:

```
procedure Flow(P: Order)
    Graph enriched by
       opns max: edge -> element
            flow: edge -> element
       eqns flow(e) =< max(e) = true
    enden
```

where P is the formal parameter and `Order` is the above theory of pre-ordered sets (called the *requirement* of the parameter). The body of the procedure, which defines what is meant by a flow in a graph, can use the sorts and operations of the parameter P by using the names (e.g. `=<`) in the requirement theory `Order`.

To supply the theory of graphs where flows are measured by natural numbers, we call the procedure with a theory of natural numbers as argument. Suppose that `NatLe` is a theory of natural numbers with sort `nat` for numbers and an operation `leq` for the usual 'less than or equal to' of natural numbers. A theory of graphs with natural numbers flows is given by:

```
Flow(NatLe[element is nat, =< is leq])
```

The list in square brackets, the 'fitting arrow', tells how the theory `NatLe` is to match the theory `Order`, that is, it defines what order we are considering on natural numbers.

The semantics of the procedure call is defined in terms of the semantics of the procedure itself and of the matching of the requirements to the actual parameters. Both of these are theory arrows. The beauty of this description is that the application of a procedure is the pushout of the two theory arrows.

Free interpretation

Theories describe a category of algebras. Often, however, we are interested not in the whole class of algebras but in a particular algebra or, more generally, some subclass. Typical cases are truth-values or natural numbers. Both of these examples are initial in a category of algebras. Thus there are occasions when we wish to interpret theories as initial or free algebras.

Consider the following theory of truth-values:

```
const Bool =
    theory data sorts bool
                 opns   true, false: bool
    endth
```

The presence of the word **data** indicates that algebras of this theory are to be initial in the category of all algebras of the given signature. In particular, the carrier has exactly two elements denoted by **true** and **false**. Extra operations on truth-values may be added by an enrichment:

```
const Bool1 =
    Bool enriched by
                 opns not: bool -> bool
                      and: bool,bool -> bool
                 eqns not(false) = true
                      not(true) = false
                      b and true = b
                      b and false = false
    enden
```

A similar example is that of natural numbers constructed from zero and a successor function by initiality:

```
const Nat =
    theory data sorts nat
                 opns   0: nat
                 succ: nat -> nat
    endth
```

Initial algebras come equipped with an equality predicate which we write as == to distinguish it from the = of equations. Thus x == y iff x = y is deducible from the equations.

When procedures are involved, it is free algebras rather than initial algebras that are described. Consider first the trivial theory of one sort:

```
const Triv =
    theory sorts element endth
```

Sequences may be described by the following procedure which is a specification of free monoids:

```
procedure Sequence(X: Triv)
    X enriched by
        data sorts sequence
            opns  empty: sequence
                  unit: element -> sequence
                  _ . _ : sequence,sequence -> sequence
            eqns  empty.s = s
                  s.empty = s
                  (s.t).u = s.(t.u)
    enden
```

The data operation is considered, like equations, to add constraints limiting the class of algebras under consideration. It is somewhat remarkable that these so-called data constraints behave, in an abstract setting, just like equations.

Derive

Often we build some complex theory and then want to extract some piece of it. As an example, instead of directly defining truth-values as above, we may want to implement them in terms of the numbers 0 and 1. Of course, some operations on numbers make no sense on truth-values so these must be omitted.

```
const Bool =
    derive sorts bool
            opns  true,false: bool
                from Nat
                by bool  is nat
                   false is 0
                   true  is succ(0)
    endde
```

The interpretation of this is a factorization of a theory arrow. It is the factorization which eliminates operations on numbers which are inappropriate for truth-values.

9.3   Theories and institutions

We now describe the model-theoretic notion of institution developed by Goguen and Burstall [1983]. The aim is to define notions of specification and theory which abstract from particular languages such as equational or first order logic. The abstract theories should be amenable to the various operations described in the previous section. This makes it possible to define the semantics of a structured specification language such as Clear without reference to any one particular base logic. Indeed, it is possible to envisage languages which use more than one base logic; for example, we might use equational logic for the definition of data types and their primitive operations and use predicate calculus to define other operations over these data types. The use of institutions for defining specification languages has become fairly popular in computer science research.

Traditionally, a logic may be described by a set S of sentences, a set M of models and a satisfaction relation $\models$ between models and sentences. However the sets S of sentences and M of models are relative to some signature, where the notion of signature varies from logic to logic. For equational logic, signatures would be as above. For predicate calculus they would also include some predicates. To define the combine and enrich operations on specifications we need a notion of inclusion between signatures and to define parameterized specifications we will need a notion of renaming between signatures. We introduce a category of signatures and must then ensure that when we change signatures we can translate sentences and models, and also that the satisfaction relation is preserved under these translations. Notice that models translate in the opposite sense from sentences.

Definition 28 *An* institution *consists of a category* **Sig** *(of objects called* signatures*) equipped with two functors and a relation* $\models$ *(called the* satisfaction *relation):*

- $Mod : \mathbf{Sig} \to \mathbf{Set}^{op}$ *– set of models on a signature,*

- $Sen : \mathbf{Sig} \to \mathbf{Set}$ *– set of sentences on a signature,*

- $\models \subseteq Mod(\Sigma) \times Sen(\Sigma)$ *for each* Σ *of* **Sig***,*

such that for each $\sigma : \Sigma \to \Sigma'$ *in* **Sig***,* $e \in Sen(\Sigma)$ *and* $m' \in Mod(\Sigma')$*,*

$$m' \models Sen(\sigma)(e) \quad \text{iff} \quad Mod(\sigma)(m') \models e$$

Moreover, we insist that the category **Sig** *is finitely cocomplete.*

An example is the *equational institution* described in Section 9.1. The category **Sig** is that of ordinary signatures (sorts and operations with their functionality), *Mod* gives the set of algebras on a signature and *Sen* the set of all equations (pairs of terms) on the signature. We might extend the notion of an institution to include a category, rather than a set, of models for each signature.

Suppose I is an institution, a *presentation* (of a theory) in I consists of a pair $(\Sigma, \mathcal{E})$ where Σ is a signature (an object of **Sig**) and $\mathcal{E}$ is a subset of $Sen(\Sigma)$.

A *theory* is a presentation $(\Sigma, \mathcal{E})$ in which the set $\mathcal{E}$ is 'closed'. In this model theory, closure is the semantic closure, defined in terms of satisfaction as follows. Extend satisfaction to sets of models M and sets of equations $\mathcal{E}$ by:

$$M \models e \quad \Longleftrightarrow \quad \forall m \in M \;\; m \models e$$
$$m \models \mathcal{E} \quad \Longleftrightarrow \quad \forall e \in \mathcal{E} \;\; m \models e$$

Given a set of sentences $\mathcal{E}$, define $M = \{m : m \models \mathcal{E}\}$ and then define the *closure* of $\mathcal{E}$ as $\mathcal{E}^* = \{e : M \models e\}$. A set of sentences $\mathcal{E}$ is *closed* iff $\mathcal{E} = \mathcal{E}^*$.

Definition 29 *A* theory *in an institution* $(\mathbf{Sig}, Mod, Sen, \models)$ *is a pair* $(\Sigma, \mathcal{E})$ *where* Σ *is an object of* **Sig** *and* $\mathcal{E}$ *is a closed subset of* $Sen(\Sigma)$.

In the equational case, the closure of a finite set of equations is, in general, an infinite non-recursive set. To deal satisfactorily with closed sets of sentences we introduce finite expressions denoting these sets. The expressions are built from the operations that we wish to perform on closed sets. The following operations are needed:

- $closure(\mathcal{E}) = \mathcal{E}^*$,

- $closeU(\mathcal{E}, \mathcal{E}') = (\mathcal{E} \cup \mathcal{E}')^*$, notice that $(\mathcal{E} \cup \mathcal{E}')^* \neq \mathcal{E}^* \cup \mathcal{E}'^*$,

- $closetrans(f, \mathcal{E}) = (Sen(f)(\mathcal{E}))^*$,

- $invtrans(f, \mathcal{E}) = \{e : Sen(f)(e) \in \mathcal{E}\}$ – it is closed,

- $star(\Sigma, \mathcal{E}) = \{e : M \models e\}$ where $M = \{m^+ : m \models \mathcal{E}\}$. It is defined when Σ has a sort 'boolean' with constants 'true' and 'false'. Then m^+ is m extended with an equality operation $==$ on each sort satisfying $a == b = true$ iff $a = b$.

The language of expressions built from these operations can be described by an ML type:

```
datatype ('o,'a,'sen)Closure =
  closure of ('sen)Set |
  closeU of ('o,'a,'sen)Closure*('o,'a,'sen)Closure |
  closetrans of 'a * ('o,'a,'sen)Closure |
  invtrans of 'a * ('o,'a,'sen)Closure |
  star of 'o * ('o,'a,'sen)Closure
```

These closure operations are the key to writing a modular theorem prover based on modular specifications. For, whilst the operations themselves are not implementable, we can use a theorem prover to check whether a sentence in an institution is in a particular closure. This idea has been used by Don Sannella [1982] to produce a pilot version of a structured theorem prover using the Edinburgh LCF system for proof construction.

Using expressions for closed sets of sentences, theories may be represented as follows:

```
datatype ('o,'a,'sen)Theory =
  theory of 'o * ('o,'a,'sen)Closure
```

A theory arrow $f : (\Sigma, \mathcal{E}) \to (\Sigma', \mathcal{E}')$ is a signature arrow $f : \Sigma \to \Sigma'$ such that $Sen(f)(\mathcal{E}) \subseteq \mathcal{E}$. Thus, including the source and target theories,

```
datatype ('o,'a,'sen)Theory_Arrow =
  theory_arrow of
    ('o,'a,'sen)Theory * 'a * ('o,'a,'sen)Theory
```

Noting that the identity and composition of theory arrows are those in **Sig**, the definition of the category of theories **Th** over an institution is straightforward. This defines a function of type:

```
cat_of_theories: ('o,'a)Cat ->
    (('o,'a,'sen)Theory,('o,'a,'sen)Theory_Arrow)Cat
```

9.4 Colimits of theories

We now show how to compute colimits of theories. We proceed by defining the initial theory, binary coproducts and coequalizers of theories and then using the general colimit construction of Chapter 4.

The initial object, binary coproducts and coequalizers of theories are defined as follows:

- The initial theory is (Φ, ϕ) where Φ is the initial signature and ϕ the empty set of equations.

- The coproduct of two theories $(\Sigma, \mathcal{E})$ and $(\Sigma', \mathcal{E}')$ is the theory $(\Sigma + \Sigma', (Sen(f)(\mathcal{E}) \cup Sen(f')(\mathcal{E}'))^*)$ where $f : \Sigma \to \Sigma + \Sigma'$ and $f' : \Sigma' \to \Sigma + \Sigma'$ is the coproduct of signatures.

- The coequalizer of $f, g : (\Sigma, \mathcal{E}) \to (\Sigma', \mathcal{E}')$ is the theory arrow $h : (\Sigma', \mathcal{E}') \to (\Sigma'', (Sen(h)(\mathcal{E}'))^*)$ where $h : \Sigma' \to \Sigma''$ is the coequalizer of f, g in the category of signatures.

As an example of a program to compute the colimit of theories, we present the coproduct function:

```
fun theory_coproduct(cSig as cocomplete_cat(Sig,_)) =
  fn (t as theory(S,E),t' as theory(S',E')) =>
    let val ((sum_sig,f,f'),univ_sig) =
            coproduct(cSig)(S,S')
        val sum_sentence =
            closeU (closetrans(f,E),closetrans(f',E'))
        val sum_theory = theory(sum_sig,sum_sentence)
        val universal =
          fn (tt,g,g') =>
          theory_arrow(sum_theory,
                       univ_sig(signature(Sig) ofo tt,
                                signature(Sig) ofa g,
                                signature(Sig) ofa g'),
                       tt) in
      ( (sum_theory,
         theory_arrow(t,f,sum_theory),
         theory_arrow(t',f',sum_theory)),
        universal )  end
```

Exercise 1. The category of theories **Th** over an institution is an indexed category, indexed by signatures. Indexed categories are defined in Chapter 5 (Exercise 12). This provides an alternative construction of colimits of theories as follows. For any institution, define the functor $F : \mathbf{Sig} \to \mathbf{Cat}$ so that $F(\Sigma)$ is the category of closed sets of Σ-sentences under inclusion. On signature arrows, $f : \Sigma \to \Sigma'$, define $(F(f))(\mathcal{E}) = (Sen(f)(\mathcal{E}))^*$. For each signature Σ, the category $F(\Sigma)$ has finite colimits (only coproducts are nontrivial: $\mathcal{E} + \mathcal{E}' = (\mathcal{E} \cup \mathcal{E}')^*$). Moreover, the functors $F(f)$ preserve finite colimits. Show that the Grothendieck construction $\mathcal{G}(F)$ is (isomorphic to) the category **Th**. Thus, by Proposition 7, **Th** has finite colimits.

9.5 Environments

A specification consists of a description of a theory in terms of other theories. Theory arrows connect the specified theory to its components. Moreover, the components themselves are defined in terms of one another. Thus the *environment* in which a theory is defined is a diagram in the category of theories. A specification denotes a theory and its connection to an environment, that is, it denotes a cocone in the category of theories. We call cocones *based objects* and their arrows *based arrows*. Based arrows are cocone arrows for which the arrow between bases is an inclusion. This restriction on the arrows expresses the way environments combine. These ideas were introduced in [Burstall, Goguen 80], a paper which should be consulted for more details.

Definition 30 *Let Δ be a diagram in a category* **C**. *Define the category* $\mathbf{C}_\Delta$ *to have, as objects, cocones whose bases are subdiagrams of Δ and, as arrows, cocone arrows which are inclusions on bases.*

Operations for combining specifications are interpreted as colimits of based objects in the category of theories. Colimits of based objects in a category **C** are constructed in terms of colimits in **C**. The construction is given explicitly in [Burstall, Goguen 80]:

Proposition 12 *If* **C** *is (finitely) cocomplete and Δ a (finite) diagram, then* $\mathbf{C}_\Delta$ *is (finitely) cocomplete.*

Proof Let δ be a (finite) diagram in $\mathbf{C}_\Delta$, with objects δ_n having apices $\hat{\delta}_n$ and bases $\beta_i \subseteq \Delta$. The colimit of δ is constructed as follows. Its base β is $\bigcup_n \beta_n$. Its apex is the colimit object of the diagram δ' which results from 'flattening' δ restricted to β; more precisely, the objects of the diagram δ' are the objects of β plus the $\hat{\delta}_n$ (i.e. the coproduct); the arrows of δ' are those of β plus the cocone arrows in each δ_n. Finally, the sides of the colimiting cocone are those resulting from the colimit of δ'. The universal property can be verified. $\square$

This construction can be programmed to give a function of type:

```
based_object_colimit: ('o,'a)CoComplete_Cat ->
    (('o,'a)Based_Obj,('o,'a)Based_Arrow)Colimit
```

The category of based objects and their colimits is then:

```
fun cocomplete_cat_of_based_objs(cC) =
      let val cocomplete_cat(C,_) = cC in
        cocomplete_cat(cat_of_based_objs(C),
                            based_object_colimit(cC)) end
```

9.6   Semantic operations

The denotation of a specification is a based object in the category of theories – naturally called a based theory. Some of the semantic operations are described in terms of colimits in the category of based theories. We describe these below.

Combine

The combine operation is the coproduct of two theories taking into account the theories shared in the environment. This is, by construction, the coproduct in the category of based theories:

```
fun combine(cSig)(b,b') =
  let val cC = cocomplete_cat_of_based_objs
                    (cocomplete_cat_of_theories(cSig))
        val ((sumbb',_,_),_) = coproduct(cC)(b,b') in
  sumbb' end
```

Enrich

The enrichment operation is a simple manipulation of signatures and sets of sentences as follows.

The theory T to be enriched is a based theory. The enrichment is a theory arrow from the signature of the apex $\hat{T}$ of T to the body of the enrichment. The resultant based theory is obtained by extending the theory arrow to $\hat{T}$ and then composing this with the sides of the cocone T.

```
fun enrich(Sig)(t,theory_arrow(_,g,theory(sig',sen'))) =
    let val th as theory(_,sen) = apex(t)
        val th' =
          theory(sig',closeU(sen',closetrans(g,sen)))
        val th_arrow = theory_arrow(th,g,th') in
    cocone_compose(cat_of_theories(Sig))(t,th_arrow) end
```

The function cocone_compose takes a cocone and an arrow out of the apex and constructs a new cocone in an obvious manner.

Procedures

The application of a procedure to its arguments is described as a pushout of based theories.

The denotation of a procedure is a based theory arrow from the co-product of the requirement theories to the body of the procedure. The application of a procedure takes this based theory arrow and a list of fitting arrows and yields the pushout based theory.

```
fun apply(cSig)(proc,fittings)
    let val cC as cocomplete_cat(bth,_) =
            cocomplete_cat_of_based_objs
                (cocomplete_cat_of_theories(cSig))
        val fa = list_coproduct(cC)(fittings)
        val ((p,q),_) = pushout(cC)(proc,fa) in
    target(bth)(p) end
```

The other specification-building operations may also be described in this categorical framework: The restriction to free interpretation is achieved by so-called data constraints which behave exactly as sentences in an institution. The 'derive' operation is a factorization of based theory arrows.

9.7 Implementing a categorical semantics

It is not the intention of this section to describe in full detail either the semantics of the specification language Clear or the implementation of this semantics undertaken by Don Sannella and the authors.

We have explained how the specification building operations are interpreted as constructions in a category. We sketch briefly how these provide a semantics for a specification language.

The abstract syntax of the language is defined as an ML type:

```
datatype Exp = just of Name
             | theory of Enr
             | plus of Exp * Exp
             | enrich of Exp * Enr
             | derive of Enr * (Exp)Set * Exp * Sic
```

```
| apply of Name * (Exp*Sic)list
| let_th of Name * Exp * Exp
```

The type `Sic` is that of signature changes, which are expressions denoting signature arrows. The type of enrichments, `Enr`, is either an ordinary enrichment (just some sorts, operations and equations) or a data enrichment.

Omitting some details concerning syntactic classes and dictionaries (for keeping track of names), we are in a position to define the semantics.

Environments are diagrams, whilst procedure environments associate a procedure name with an arrow of based theories and a list of based theories – the requirements.

The semantic function gives the denotation of expressions as based theories.

```
E: Exp -> (('o,'a)CoComplete_Cat ->
(('o,'a)Env -> (('o,'a)ProcEnv -> ('o,'a)Based_Theory)))
```

For example, the expression for combining two theories is defined by:

```
fun E(plus(e1,e2))(cSig)(rho)(pi) =
combine(cSig)(E(e1)(cSig)(rho)(pi),E(e2)(cSig)(rho)(pi))
```

The other expressions are handled in a similar style, invoking the appropriate categorical constructions. Full details of the semantics may be found in [Burstall, Goguen 80]. For descriptions of this implementation consult [Sannella 82] or [Rydeheard 81].

As a practical implementation of a specification language, this is not satisfactory. It is a large program which runs slowly. For the sake of comparison and to get a practical program, Don Sannella [1984] has written a direct implementation of a variant of this semantics. The semantic operations, instead of being interpreted in a general category, are specialized to operations on equational theories and are given directly in set theory. This speeds up the the program considerably, suggesting that a source of inefficiency is the use of general high-level code. It also suggests that this improvement may be achievable through program transformation but this is not clear. However, it seems that much of the improvement is due to the reduced size of the programs and data structures in the ML system.

Chapter 10

Formal Systems for Category Theory

The advent of computing has given renewed interest in formal systems for describing areas of mathematics. The interest is not simply meta-mathematical – establishing properties of formal systems – but practical. We actually want to formalize mathematics for machine processing.

We have seen in previous chapters how constructions from category theory can be expressed in the programming language ML. In this chapter we consider formal (linguistic) aspects of category theory. We list some of the requirements on a formalism for category theory. Some of these we have come across as features of ML, others arose where ML proved inadequate for programming category theory.

We also display fragments of category theory written in some machine-implemented languages other than ML. This is not meant to be an exhaustive account but merely to indicate possibilities. We consider two languages: OBJ, a language of theories and algebras, and GTTS, an implementation of a constructive type theory. We also look at Hagino's [1987] system in which universal constructs in category theory are used as a general type definition mechanism. We are indebted to J. A. Goguen for allowing us to present his OBJ program and to R. Dyckhoff for allowing us to report his formulation of category theory in a variant of Martin-Löf's type theory. The latter has appeared as [Dyckhoff 85].

An interesting use of cartesian closed categories in programming occurs in the Categorical Abstract Machine. We do not describe this here; a full account may be found in [Curien 86].

10.1   Formal aspects of category theory

Category theory provides a language for elementary (first order) descriptions of some general phenomena in mathematics. Mathematicians have been interested in this linguistic aspect of category theory, partly from a foundational point of view so as to avoid paradoxes, and partly to turn the notation used in practice into a formal framework, see [Bénabou 85]. Our interest is in appropriate languages in which to embed category theory for machine processing.

Here we list some formal aspects of category theory:

- Categories are algebraic – they can be presented in terms of operations and equations. This applies also to categorical concepts such as products, coproducts, and equalizers.

  The operations may be only partially defined, for example composition is defined not on all pairs of arrows but only on composable pairs. This is called 'equational partialness', or 'essentially algebraic' [Freyd 72], in that operations can be arranged in order so that the domain of definition of each partial operation is an equation in preceding operations. Equational partialness has been investigated by Cartmell [1978], Benecke and Reichel [1982] and Burmeister [1986]. The somewhat more general notion of 'left exact theories' is discussed in [Barr, Wells 85] and [McLarty 86]. Lambek and Scott [1986] point out that categories are graph algebras – algebras whose carriers are graphs rather than sets. There are also formal systems for handling more general kinds of partial functions. These include order-sorted algebras [Goguen 78], conditional rewriting systems [Bergstra, Klop 82], multi-valued logics [Barringer et al. 84], definedness operators [Scott 79] and [Beeson 86], the logic of computable functions, LCF, [Gordon et al. 79], domain-theoretic methods [Plotkin 1985] and categorical approaches to partial functions [Robinson, Rosolini 86].

- Commutative diagrams abound in category theory. Categorical reasoning in the form of 'diagram chasing' is typed equational deduction as pointed out by Huet [1986] and is open to some form of automation [Watjen, Struckmann 82]. The diagrammatic nature of category theory suggests the use of programming languages more pictorial than current languages.

- We often consider categories whose class of objects encompasses all algebraic structures of a certain kind, for example, varieties of algebras or categories of functors. These categories are second order structures. Formalisms based on theories and their algebras, such as program modules, are inherently first order and cannot cope with such categories. Though modules have a type (their 'interface' – a signature or theory), this cannot appear as a type within a module.

- Category theory supports a principle of duality. The theory is invariant under 'reversing arrows'. This may be formalized at a meta-level as a truth-preserving transformation of formulae in category theory. Thus for each proof, we obtain two theorems, that proven and its dual. Duality may be expressed as an operation on categories, taking a category to that with its arrows reversed. This may be programmed and allows us to obtain two calculations from a single program through composition with duality operations.

- Infinite structures naturally arise in category theory, even if we restrict ourselves to computational aspects of the theory. An example is the construction of term algebras on finite sets. We may introduce recursive or r.e. sets to accommodate such constructions but it often seems more natural to describe these infinite sets as types rather than values. For instance, we want the type of lists, as well as that of integers, to be objects in the category of monoids. This 'objects as types' is especially necessary when dealing with term algebras as they introduce new definitional mechanisms for functions. By moving to categories in a higher universe we may let objects range over types as required.

- There is a correspondence between typed functional programming and category theory. Types correspond to objects, functions are arrows and type constructors are functors. However, functions, in the presence of possible non-termination, are partial functions, whereas arrows are usually taken to be total (e.g. in **Set**).

- Reflection principles and predicativity. Category theory is unusual amongst mathematical theories in that categories are objects in a category whose arrows are functors. Most type theories are predicative in character and cannot comfortably handle such reflection. For instance in [Martin-Löf 84] there is a hierarchy of type universes. We can introduce categories whose objects have types ranging over a universe. However, categorical constructions are coded for each universe and cannot be applied to higher universes even though the construction is identical for each universe. Attempts to work within one universe can lead to inconsistency, as discussed by Coquand [1986].

• Categories like **Set** and **FinSet** are non-homogeneous structures in that the type of elements in the sets varies with the set. ML polymorphism is not sufficient to deal with these structures. If only homogeneous categories of sets are considered then operations on objects impose requirements on the type of the elements. For example, the disjoint union of sets constructed as a labelled union requires the labelled elements to be of the same type as the original elements.

• We have emphasized that universal constructs in category theory can be expressed as higher order functions. Higher order structure abounds in category theory, for example, the the action of the free monoid functor on arrows is the 'maplist' function ('mapcar' in Lisp) defined in ML by

```
fun maplist(f)(nil) = nil
  | maplist(f)(x::s) = f(x)::maplist(f)(s)
```

Goguen, in the OBJ example below, shows how some of these higher order functions can be treated in a first order framework.

• Equality on objects. When working in category theory we do not usually make statements like 'for objects a and b, $a = b$'. We consider isomorphism rather than equality on objects. However, the definition of a category requires that composition is defined for pairs g, f such that the source of f is equal to the target of g – an equality on objects. We came across this matter in Chapter 3 when writing programs for the composition function. It is discussed in [Bénabou 85], where a distinction is drawn between what is expressible in category theory and what is expressible in a meta-language in which we define categories.

We now turn to category theory expressed in several quite different formalisms.

10.2 Category theory in OBJ

OBJ is a language based upon on theories and algebras. Theories serve as specifications of intended behaviour whilst algebras are implementations of the theories. Algebras are called 'objects' in OBJ. Beware of confusion with objects in categories. OBJ was devised by J. A. Goguen and first implemented by J. Tardo [Goguen, Tardo 79]. We use a version of OBJ3 [Goguen 88], which developed from OBJ2 [Futatsugi et al. 85].

Theories and algebras are presented as collections of operations together with axioms which are either equations or conditional equations. OBJ is a programming language in which evaluation is equational rewriting using the axioms in algebras. Theories and algebras are linked by 'views' which assert that an algebra is considered as belonging to a given theory. A view is a renaming map, mapping names of sorts and operations in the theory to those corresponding in the algebra.

Let us have a look at a simple example of theories and algebras so as to fix notation. In the example below we define the theory (introduced by th) of monoids (semigroups with identity). We introduce a particular monoid as an OBJ object (introduced by obj) and assert through a view that the object is an algebra of the theory of monoids.

```
th MONOID is
  sort El .
  op _;_ : El El -> El [assoc] .
  op E : -> El .
  var A : El .
  eq : A ; E = A .
  eq : E ; A = A .
endth

obj INTS is
  protecting INT .
  op _++_ : Int Int -> Int [assoc] .
  op two : -> Int .
  vars A B : Int .
  eq : A ++ B = A * B - A - B + 2 .
  eq : two = 2 .
endo

view INTS-AS-MONOID from MONOID to INTS is
  sort El to Int .
```

```
op : _;_ to : _++_ .
op : E to : two .
endv
```

Equational rewriting in OBJ makes special provision for some axioms like associativity, commutativity, idempotence and identity. These are treated as attributes of operations and special routines are built-in for matching and unification in the presence of these axioms. This is illustrated above where the binary operation in a monoid is associative. It also has an identity which we could treat in the same way but instead make into an operation satisfying appropriate equations, so as to illustrate equations in a theory. Denotationally these are the same, but operationally the equations serve as left to right rewrite rules, whereas the specialist matching algorithm for identity will match in both directions. Equations may be conditional, in which case the condition, an equation, is evaluated to determine whether or not the rewrite rule is applicable.

OBJ allows operations to be partial (indicated by 'op_as') with the domain of definition given by equations in preceding operations. This form of partialness goes under various names including 'equational partialness' [Burmeister 86], 'essentially algebraic' [Freyd 72] and the somewhat more general 'left exact' [Barr, Wells 85].

In OBJ, the sorts may be partially ordered so that one sort S is a subsort of another S', written S < S'. When this occurs, operations on S' may be applied to values of sort S, whereas operations returning values of sort S may be considered to return values of sort S'.

Theories and algebras may be built using already defined theories and algebras. In the example above we build the monoid INTS from that of integers INT but we wish to protect integers from modification (e.g. by adding an equation like 1 = 2). Structures may be parameterized using a parameterization technique similar to that of Clear discussed in Chapter 9. As an example, we consider an algebra of lists:

```
obj LIST[X :: TRIV] is
  sorts List .
  subsorts Elt < List .
  op _._ : List List -> List [assoc, id : nil] .
endo
```

This is parametrized on the type of elements in the list. Their are no requirements on the elements to form lists so the formal parameter X is matched to the trivial theory TRIV having just one sort Elt. We use the subsort mechanism to assert that elements are singleton lists. Lists of, say, integers are obtained by instantiating with a view showing how integers may be considered as algebras of TRIV:

```
view INT-AS-TRIV from TRIV to INT is
  sort Elt to Int .
endv
```

Then LIST[INT-AS-TRIV] denotes the OBJ object of lists of integers. Non-trivial requirements on parameters occur, for example, in sorting where we require the elements of lists to support an ordering, so must match the theory of partial (or total) orders.

Because categories are algebras and categorical concepts tend to be algebraic in nature, a language like OBJ would appear to be a useful formalism for category theory. Indeed, the equational partialness and the special treatment of associative operations are just what is required for the composition of arrows. Moreover, the order which may be imposed upon sorts allows us to express the conversion between types when one type is extended by extra components or extra properties. An example in category theory is the fact that colimiting cocones are cocones together with their universality so we may write Colimit-CoCone < CoCone. Structuring facilities, like those in OBJ, are essential in building large and complex systems, both to control the complexity and to allow sufficient generality.

The framework of theories and algebras adopted by OBJ is first order. As already remarked, categories such as varieties of algebras and categories of functors are second order structures in that their objects range over all structures of a certain kind. Thus whilst we can define the theory of categories and the theory of monoids in OBJ, we cannot match the sort of objects in categories to the theory of monoids to form the category of monoids. This may be circumvented but is a mismatch between the OBJ treatment and mathematical practice.

We now turn to a fragment of category theory in OBJ3 kindly supplied by J. A. Goguen. We begin with the theory of categories. The associativity of composition and its partial nature are both directly expressible:

```
th CAT-TH is
  sorts Obj Arrow.
  op d0_ : Arrow -> Obj .
  op d1_ : Arrow -> Obj .
```

```
    op-as _;_ : Arrow Arrow -> Arrow
        for M1 ; M2 if d1 M1 == d0 M2 [assoc] .
    op id_ : Obj -> Arrow .
    var O : Obj .
    vars A A0 A1 : Arrow .
    eq : d0 id O = O .
    eq : d1 id O = O .
    eq : d0 (A0 ; A1) = d0 A0 .
    eq : d1 (A0 ; A1) = d1 A1 .
    eq : (id d0 A) ; A = A .
    eq : A ; id d1 A = A .
  endth
```

Categories of sets form algebras (OBJ objects) of this theory. The categories of sets which we define are homogeneous, in that elements of every set in a category have the same type. We first define these categories and then define sets.

```
  obj CAT-SET[X :: TRIV] is
    protecting SET[X] .
    sorts Fn .
    op d0_ : Fn -> Set .
    op d1_ : Fn -> Set .
    op-as _;_ : Fn Fn -> Fn
        for F1 ; F2 if d1 F1 == d0 F2 [assoc] .
    op id_ : Set -> Fn .
    op-as _of_ : Fn Elt -> Elt
        for F of X if (X in d0 F) and (F of X in d1 F) .
    var O : Set .
    vars F F0 F1 : Fn .
    var E : Elt .
    eq : d0 id O = O .
    eq : d1 id O = O .
    eq : d0 (F0 ; F1) = d0 F0 .
    eq : d1 (F0 ; F1) = d1 F1 .
    eq : (id d0 F) ; F = F .
    eq : F ; id d1 F = F .
    eq : (F0 ; F1) of E = F0 of (F1 of E) .
    eq : (id O) of E = E .
  endo
```

The definition is parameterized over the type of elements in the sets.

The elements require no structure, so need only match the theory TRIV. Objects in the category are sets imported from SET[X] which we define below. Arrows are defined as functions with a source set and a target set. The functional behaviour of arrows is given by the application operation 'of'. The last two equations define composition and identity in terms of application. Moreover, arrows are total functions as asserted in the condition for definedness of application.

A view is used to assert that this OBJ object matches the theory of categories:

```
view CAT-SET-AS-CAT from CAT-TH to CAT-SET is
  sort Obj to Set .
  sort Arrow to Fn .
endv
```

Sets are usually defined in terms of union and intersection. Instead, to simplify the axiomatic treatment, we make intersection and symmetric difference the primary operations, defining union in terms of them. That is, we introduce the Boolean ring of sets rather than the Boolean algebra. A Boolean ring is a ring in which every element satisfies $A.A = A$. A well-known result of algebra asserts that these two algebraic structures are equivalent. For sets, multiplication is intersection and addition is symmetric difference. We introduce a top element (a universal set Ω) as well as a bottom element (the empty set) so that sets constructed from the operations are either finite or formally cofinite, i.e. of the form $\Omega - S$ for finite S.

```
obj BSET[X :: TRIV] is
  protecting BOOL .
  sort Set .
  op {} : -> Set .
  op omega : -> Set .   *** universal set
  op {_} : Elt -> Set .
  op _+_ : Set Set -> Set [assoc comm id: {}] .
                    *** symmetric difference
  op _&_ : Set Set -> Set [assoc comm idem id: omega] .
                    *** intersection
  vars S S' S'' : Set .
  vars E E' : Elt .
  eq : S + S = {} .
  ceq : { E } & { E' } = {} if E =/= E' .
```

```
  eq : S & {} = {} .
  eq : S &(S' + S'') = (S & S')+(S & S'') .
endo
```

Other standard operations on sets are defined as follows:

```
obj SET[X :: TRIV] is
  protecting BSET[X] .
  protecting INT .
  op _U_ : Set Set -> Set [assoc comm id: {}] .
  op _-_ : Set Set -> Set .
  op #_ : Set -> Int [prec 0] .
  op _in_ : Elt Set -> Bool .
  op _in_ : Set Set -> Bool .
  op empty?_ : Set -> Bool .
  var X : Elt .
  vars S S' S'' : Set .
  eq : S U S' = (S & S')+ S + S' .
  eq : S - S' = S + (S & S') .
  eq : empty? S = S == {} .
  eq : X in S = { X } & S =/= {} .
  eq : S in S' = S U S' == S' .
  eq : # {} = 0 .
  ceq : #({ X } + S) = # S if X in S .
  ceq : #({ X } + S) = 1 + # S if not X in S .
endo
```

Let us look at binary coproducts in this formalism. We start with binary cocones in a category. The OBJ code below describes cocones as a pair of arrows with common target. It is parameterized on categories and uses an OBJ object of pairs for bases of cocones.

```
obj CO2CONE[C :: CAT-TH] is
  protecting 2TUPLE[Obj,Obj] * (sort 2Tuple to Base) .
  sorts Co2cone .
  op-as cocone : Arrow Arrow -> Co2cone
      for cocone(A1,A2) if d1 A1 == d1 A2 .
  op j1 : Co2cone -> Arrow .
  op j2 : Co2cone -> Arrow .
  op apex : Co2cone -> Obj .
  op base : Co2cone -> Base .
  vars A1 A2 : Arrow .
```

```
    eq : j1(cocone(A1,A2)) = A1 .
    eq : j2(cocone(A1,A2)) = A2 .
    eq : apex(cocone(A1,A2)) = d1 A1 .
    eq : base(cocone(A1,A2)) = << d0 A1 ; d0 A2 >> .
  endo
```

Categories with coproducts form a theory in OBJ. An algebra of the theory is a category together with a construction of coproducts of pairs of objects. Coproducts, as universal cocones, are a subsort of cocones and cocone operations are inherited implicitly. The universality is expressed as an operation which, for a universal cocone and a cocone on the same base, returns the mediating arrow. Unlike universality as expressed in ML, this gives a first order treatment by separating components and using sorts and equations to link them. Notice that OBJ is really second order in functionality as names are given to operations in theories. These names may be bound to actual operations using a view.

```
th CO2PROD-TH[C :: CAT-TH] is
  protecting CO2CONE[C] .
  protecting BOOL .
  sort Uco2cone .
  subsorts Uco2cone < Co2cone .
  op ucocone : Obj Obj ->Uco2cone . ***coproduct cocone
  op _++_ : Obj Obj -> Obj . *** coproduct object
  op-as uarrow : Uco2cone Co2cone -> Arrow
        for uarrow(U,C) if base(U) == base(C) .
  vars A B : Obj .
  vars F G H : Arrow .
  eq: apex(ucocone(A,B)) = A ++ B .
  eq: base(ucocone(A,B)) = << A ; B >> .
  eq: (j1(ucocone(A,B)));uarrow(ucocone(A,B),cocone(F,G))
        = F .
  eq: (j2(ucocone(A,B)));uarrow(ucocone(A,B),cocone(F,G))
        = G .
  ceq: H = uarrow(ucocone(A,B),cocone(F,G))
          if (j1(ucocone(A,B)); H == F)
            and (j2(ucocone(A,B)); H == G) .
endth
```

We can now give a calculation of coproducts in the categories of sets defined above. Coproducts of sets are disjoint unions. Disjoint unions may be expressed in various ways according to the type of elements in the sets. We abstract on this calculation by introducing a theory of injection arrows into coproducts. The description of coproducts of sets is parameterized on this theory. This is then specialized to sets of integers using an arithmetic formula for the disjoint union.

```
th 2INJ-TH is
   protecting BOOL .
   sort Elt .
   op i0 : Elt -> Elt .
   op i0inv : Elt -> Elt .
   op i0pred : Elt -> Bool .
   op i1 : Elt -> Elt .
   op i1inv : Elt -> Elt .
   op i1pred : Elt -> Bool .
   var E : Elt .
   eq : i0inv(i0(E)) = E .
   eq : i1inv(i1(E)) = E .
   eq : i0pred(i0(E)) = true .
   eq : i0pred(i1(E)) = false .
   eq : i1pred(i1(E)) = true .
   eq : i1pred(i0(E)) = false .
endth

obj CO2PROD-CAT-SET[X :: 2INJ-TH] is
   extending CO2CONE[CAT-SET[X]] .
   sort Uco2cone .
   subsorts Uco2cone < Co2cone .
   op ucocone : Set Set -> Uco2cone .
   op-as uarrow : Uco2cone Co2cone -> Fn
        for uarrow(U,C) if base(U) == base(C) .
   op I0 : Set -> Set .
   op I1 : Set -> Set .
   op _++_ : Set Set -> Set .
   vars A B S : Set .
   vars F G : Fn .
   var E : Elt .
   eq : I0({}) = {} .
   eq : I0({ E } + S) = { i0(E) } + I0(S) .
   eq : I1({}) = {} .
   eq : I1({ E } + S) = { i1(E) } + I1(S) .
```

```
eq : A ++ B = I0(A) U I1(B) .
eq : apex(ucocone(A,B)) = A ++ B .
eq : base(ucocone(A,B)) = << A ; B >> .
ceq : j1(ucocone(A,B)) of E = i0(E) if E in A .
ceq : j2(ucocone(A,B)) of E = i1(E) if E in B .
ceq : uarrow(ucocone(A,B),cocone(F,G)) of E =
                  F of i0inv(E) if i0pred(E) .
ceq : uarrow(ucocone(A,B),cocone(F,G)) of E =
                  G of i1inv(E) if i1pred(E) .
endo


view CO2PROD-CAT-SET-AS-CO2PROD-TH[J :: 2INJ-TH]
   from CO2PROD-TH[CAT-SET[J]] to CO2PROD-CAT-SET[J] endv
```

The category of sets of integers with an explicit calculation of co-products is defined below. A view is used to instantiate the theory of coproducts to the coproduct injections calculated numerically.

```
obj CAT-SET-INT is
  protecting CAT-SET[INT] * (op omega to ints) .
endo

obj CO2PROD-CAT-SET-INT is
  protecting CO2PROD-CAT-SET[view from 2INJ-TH to INT is
    sort Elt to Int .
    var I : Elt .
    op : i0(I) to : (2 * I) .
    op : i0inv(I) to : (I quo 2) .
    op : i0pred(I) to : (I rem 2 == 0) .
    op : i1(I) to : 1 + (2 * I) .
    op : i1inv(I) to : ((I - 1) quo 2) .
    op : i1pred(I) to : (I rem 2 == 1) .
  endv] .
endo

view CO2PROD-CAT-SET-INT-VIEW
        from CO2PROD-TH[CAT-SET-INT]
        to CO2PROD-CAT-SET-INT endv
```

10.3   Category theory in a type theory

We now look at a fragment of category theory in a type theory for constructive mathematics. Type theories are organized around types and values and the notion that a value has a particular type. Usually, type constructors include dependent types and some form of type universes as well as product and function types. Rules are given by which we may establish whether a given expression is a type and whether an expression for a value belongs to a particular type. Unlike type systems in programming languages, these theories allow types to be defined with axiomatic constraints. Type checking is often not decidable. Programs appear in this framework as constructive proofs. This is to be contrasted with OBJ where both types (sets and operations on them) and algorithms (rewrite rules) are identified in the concept of an OBJ object.

A variety of constructive type theories have been proposed including those of Martin-Löf [1975,82,84] and Feferman [1975,79], the Theory of Constructions [Coquand, Huet 85], the 'logical' theories of [Aczel 80] and [Dybjer 85] and the Logical Framework (LF) [Harper et al. 87]. Automated systems have been developed such as Nuprl [Constable et al. 85] and the Göteborg Type Theory System [Petersson 84], both based on Martin-Löf's work, and the Automath system [de Bruijn 80]. The Theory of Constructions and the Logical Framework are also available as machine implementations.

We summarize the work of Roy Dyckhoff [1985] in developing an experimental implementation of category theory in the Göteborg Type Theory System (GTTS) which implements a type theory of Martin-Löf. Inference rules are presented in the natural deduction style, so are organized into those that introduce operators and those that eliminate them. Moreover there are rules for establishing equality of types and of values.

To implement category theory in such a system we may represent the types (like that of categories) and values (like **FinSet**) in terms of those provided by the system and hence use the inference rules of the system itself. Categories are introduced axiomatically and it is necessary to prove that such an axiomatic presentation denotes a type of the system. This approach is logically sound but leads to clumsy notation and fails to capture the linguistic content of category theory in providing a typed arrow-theoretic language.

There is an alternative. Expand the type system with new rules for type formation (for the type of categories, functors and other categorical concepts such as natural transformations) and value formation (such as the composition of arrows). Adding new rules may violate consistency, so a meta-level proof of consistency is required.

To express the fact that categories form a type we use the simple rule of type formation in which the requirements for forming the type (above the horizontal line of the rule) are empty:

$$\overline{CAT \ type}$$

Strictly speaking, to avoid inconsistency, we should declare CAT to be a type within a universe U_1 ($CAT : U_1$) and the subsequent rules should be amended accordingly, replacing *type* with U_1.

Declaring functors to be a type requires two categories. We form the type of all functors between two categories as follows:

$$\frac{C : CAT \quad D : CAT}{FUNC(C, D) \ type}$$

We read this as 'from the assertion that C and D are categories, infer that $FUNC(C, D)$ is a type'.

To introduce a value of type CAT we need some ingredients: a type of objects, a family of types of arrows; two functions, identity and composition; and some axioms. This is the introduction rule for categories:

$$\frac{\begin{array}{l} O \ type \\ A(X, Y) \ type \quad [X : O, Y : O] \\ i : A(X, X) \quad [X : O] \\ c(f, g) : A(X, Z) \quad [X : O, Y : O, Z : O, f : A(X, Y), g : A(Y, Z)] \\ c(f, i) = f : A(X, Y) \quad [X : O, Y : O, f : A(X, Y)] \\ c(i, f) = f : A(X, Y) \quad [X : O, Y : O, f : A(X, Y)] \\ c(f, c(g, h)) = c(c(f, g), h) : A(W, Z) \\ \quad [W : O, X : O, Y : O, Z : O, f : A(W, X), g : A(X, Y), h : A(Y, Z)] \end{array}}{Cat(O, A, i, c) : CAT}$$

The lists in square brackets are the assumptions for the assertion to the left, i.e. e [e'] is the assertion that e holds under the assumptions e'. Notice how dependent types handle the source and target of arrows and the partial nature of composition. Notice also the presence of axioms, showing that we need rules for establishing equality of arrows. There is a rule asserting that any category may be constructed as above. Elimination rules allow us to extract components of categories ('selectors'); Ob and $Arrow$ for the types of objects and arrows, and *comp* and *id* for the composition and identity functions.

The rule for introduction of functors looks like:

$$C : CAT$$
$$F1(X) : Ob(D) \quad [X : Ob(C)]$$
$$F2(f) : Arrow(D, F1(X), F1(Y))$$
$$[X, Y : Ob(C), f : Arrow(C, X, Y)]$$
$$F2(id) = id : Arrow(D, F1(X), F1(X)) \quad [X : Ob(C)]$$
$$F2(comp(f, g)) = comp(F2(f), F2(g)) : Arrow(D, F1(X), F1(Z))$$
$$[X, Y, Z : Ob(C), f : Arrow(C, X, Y), g : Arrow(C, Y, Z)]$$

$$\overline{Func(F1, F2) : FUNC(C, D)}$$

Dyckhoff presents rules for natural transformations and has implemented a facility for reducing expressions for arrows using associativity and identity laws. This fragment of category theory is presented with some 40 inference rules. He also presents some proofs in this system including the reflexivity of isomorphism and the fact that the composite of monics is a monic.

Particular categories are defined using the introduction rule for values of type *CAT*. We may form a category of small types, i.e. types in the universe U_1, whose arrows from X to Y are values of type $X \rightarrow Y$. This category is cartesian closed. Other categories such as categories of functors and categories of categories can be defined within this framework. For these we need to consider type universes and we get into the hierarchical situation described earlier. The reflective nature of category theory is not accommodated in this type theory.

10.4 Categorical data types

A very interesting implementation of some categorical concepts has been undertaken by Tatsuya Hagino [1987,87a] at Edinburgh University. He defines a programming language whose sole declaration mechanism is essentially definition of a left or right adjoint functor. With this minimal apparatus he is able to define the various features available in conventional functional languages such as ML, for example, products, sums, exponentials, data types and lazy (i.e. infinitary) data types. Thus his use of category theory is essentially to define a programming language rather than to define particular programs. In fact, his work can be viewed as a 'rational reconstruction' of a prototypical functional programming language on categorical principles.

It is unclear whether his methods could be extended to cover the area treated in this book. He works relative to a single underlying category, **U**, which is uninterpreted but may be thought of as **Set**. Types are represented by objects, parameterized types by functors, functions by arrows and data values by arrows from the terminal object. This differs from our approach in which values in ML correspond to objects and perhaps seems more natural.

Each declaration is introduced with the key words `right object` or `left object` and defines a type and some associated functions, i.e. a functor and associated unit or co-unit natural transformations plus a bijection between hom-sets. Some of the flavour of the system may be obtained from the following examples.

Products are declared to be a right adjoint in the following rather succinct manner:

```
right object prod(t1,t2) with pair is
   pi1: prod -> t1
   pi2: prod -> t2
end object
```

In this definition a new functor `prod` is defined together with the co-unit, the pair of arrows `pi1` and `pi2` and the bijection `pair` : t1 × t2 → prod(t1, t2). The new functor is right adjoint to a functor which is determined implicitly by the expressions after `pi1` and `pi2`; this is the diagonal functor from **U** to **U** × **U**. The pair of arrows correspond to the co-unit of the adjunction.

In fact, the definition mechanism must be a little more general than adjoints, and it is based on parameterized F,G-dialgebras. First, a T-algebra over **C**, where $T : \mathbf{C} \to \mathbf{C}$ is a functor, is an object c and an arrow $f : T(c) \to c$. Extending this notion, given functors $F, G : \mathbf{C} \to \mathbf{D}$, Hagino defines an F,G-dialgebra as an object c together with an arrow $h : F(c) \to G(c)$. Left and right objects are defined in terms of initial and final F,G-algebras.

Definition 31 *Let $F, G : \mathbf{C} \to \mathbf{D}$ be functors, the category* $\mathbf{DAlg}(F, G)$ *has objects pairs $\langle c, f \rangle$ where c is a **C**-object and $f : F(c) \to G(c)$ in* **D**, *arrows from $\langle c_1, f_1 \rangle$ to $\langle c_2, f_2 \rangle$ are arrows $h : c_1 \to c_2$ such that* $G(h)f_1 = f_2 F(h)$.

More precisely, Hagino uses parameterized F,G-dialgebras to deal with parameterized data types. Consider functors $H, K : \mathbf{A} \times \mathbf{B} \to \mathbf{D}$, and write $H_a(b)$ for $H(a, b)$ and $K_a(b)$ for $K(a, b)$. Then for each a in **A**, let $\langle Left(a), \eta_a \rangle$ be the initial algebra of $\mathbf{DAlg}(H_a, K_a)$, if it exists. We can extend *Left* to a functor from **A** to **B**, and $\eta_a : H_a \to K_a$ is a natural transformation. Dually, we get a functor $Right : \mathbf{A} \to \mathbf{B}$ and a natural transformation ϵ by taking the final algebra, if it exists.

Given a functor $F : \mathbf{A} \to \mathbf{D}$, we can put $H_a(b) = b$ and $K_a(b) = F(a)$; Hagino shows that in this case *Left* is the left adjoint of F. Thus the construction really does generalize the notion of adjunction. Interestingly, it also yields comma categories as a specialization.

In a left object definition we supply H and K and hence define the functor $Left_{\langle H,K \rangle}$ and the natural transformation η as well as the map from an arbitrary algebra to the unique dialgebra homomorphism to it. This map ϕ corresponds to the bijection of hom-sets in an adjunction. We illustrate this with the example of coproduct.

```
left object coprod(t1,t2) with case is
    in1: t1 -> coprod
    in2: t2 -> coprod
end object
```

Here the given information is:

- $\mathbf{A}$ is $\mathbf{U} \times \mathbf{U}$ – because there are two parameters, t1 and t2

- $\mathbf{B}$ is $\mathbf{U}$ – this is always the case

- $\mathbf{D}$ is $\mathbf{U} \to \mathbf{U}$ – because there are two arrows in1 and in2

- $H_{(\mathbf{t1,t2})}(\mathbf{coprod})$ is $(\mathbf{t1,t2})$ – source of $(\mathbf{in1,in2})$

- $K_{(\mathbf{t1,t2})}(\mathbf{coprod})$ is $(\mathbf{coprod,coprod})$ – target of $(\mathbf{in1,in2})$

and the results of the declaration are:

- *Left* is $\mathbf{coprod} : \mathbf{U} \times \mathbf{U} \to \mathbf{U}$ – the new functor, a type constructor

- $\eta_{(\mathbf{t1,t2})}$ is $(\mathbf{in1,in2})$ – the injection arrows

- ϕ is $\mathbf{case}$ – if $f_1 : \mathbf{t1} \to \mathbf{t}$ and $f_2 : \mathbf{t2} \to \mathbf{t}$ then $\mathbf{case}(f_1, f_2) :$ $\mathbf{coprod(t1,t2)} \to \mathbf{t}$

Exponentials are right adjoints defined in terms of products. As well as the new type, a `curry` function and an `eval` function are defined.

```
right object exp(s,t) with curry is
    eval: prod(exp,s) -> t
end object
```

Lists are defined in terms of products as follows. The definition is very like an ML data type definition, but it also delivers a functional to do primitive recursion on lists.

```
left object list(t) with primitive_list_rec is
  nil: 1 -> list
  cons: prod(t,list) -> list
end object
```

A dual definition defines 'lazy' lists, that is infinite lists.

Functions over the data types constructed in this way can be defined from the primitives provided by the left and right object definitions. Hagino uses composition rather than lambda abstraction to do this, rather in the style of Backus' FP language. He first presents his style of definition as a specification language, an alternative to algebraic specification languages. This requires a precise description of well-formed functorial expressions. He then shows that if the underlying category has certain properties, notably cartesian closure, his definitions yield a Categorical Programming Language. The equations satisfied by the unit (or co-unit), and by the bijection part of an adjunction generalized as shown above, can be used as rewrite rules. This gives a evaluation algorithm, which he has implemented to build an interpreter for his language in Franz Lisp. Using Tait's method, he proves that the computations always terminate, a substantial theoretical result given the power of his language.

The remarkable aspect of Hagino's Categorical Programming Language is the economy of means with which he is able to define the various facilities provided by the usual style of functional programming language. Making use of one categorical definition mechanism, he no longer has to take the terminal object, products, coproducts or exponentials as primitive notions. He defines data types and lazy data types in a dual manner. It should be noted that he is defining a categorical programming language, whereas in this book we have been using a traditional functional programming language to code categorical constructions.

Appendix A

ML Keywords

The table below contains ML keywords together with a reference to the page on which the ML construct is described.

Keyword	Page number
abstype	25
and	12
as	21
case	22
datatype	21
exception	26
fn	13
fun	13
handle	26
if then else	12
infix	13
let	12
of	22
raise	26
type	17
val	11
:	16
*, (,)	16
->	13
;	12
|	21
,	23

Appendix B

Index of ML Functions

Appendix C

Other ML Functions

This is a list of ML functions which are used in the text, excluding those in Appendix B. In the main, these are simple functions and, to avoid interrupting the text, their action is described here.

applyda   Applies a diagram of functors to an arrow, returns a diagram arrow

applydo   Applies a diagram of functors to an object, returns a diagram

apply_Fun_Diag   Applies a functor to a diagram

apply_Fun_CoCone   Applies a functor to a cocone

apply_Nat_Diag   Applies a natural transformation to a diagram, returns a diagram arrow

a_prod_o_within   In a complete category, the product of an arrow with the identity arrow on an object

as_functor   The functor obtained by fixing an argument of a bifunctor

base   Extracts the base diagram of a cocone

co_apex   Extracts the apex of a cocone

co_apex_arrow   Extracts the arrow between the apices of a cocone arrow

compose   Takes a category, returns its composition of arrows

coproduct   Takes a cocomplete category, returns a binary coproduct function

domain   The source category of a functor

fn_to_list   Converts finite functions to association lists (their graphs)

Fun_comp   The composition of functors

identity   Takes a category, returns the function assigning identity arrows to objects

list_to_fn   Converts association lists (graphs of functions) into functions

mapset   The image of a set through a function

new_cocone   In a category, takes cocone $\xi_n : \Delta'_n \to a$ on diagram Δ' and a diagram arrow $\delta : \Delta \to \Delta'$, returns the cocone $\xi_n \delta_n : \Delta_n \to a$

obj_at_node   Takes a diagram, returns the object at a given node

OF   Applies an arrow in **FinSet** to an element of the source set

ofa   Applies a functor to an arrow

ofa   Applies a functor to an object

product   Takes a complete category, returns a product function

pushout   Takes a cocomplete category, returns a pushout function

pullback   Takes a complete category, returns a pullback function

range   The target category of a functor

restrict   The restriction of an arrow in **FinSet** to a subset of the source

sides   Takes a cocone, returns the function from nodes in the base to arrows to the apex

signature   Given a category of signatures (in an institution), the functor returning the signature of a theory

sharp   Takes an adjunction (F, G, η, ϵ), returns the function $(f : a \to G(b)) \mapsto (f^\# : F(a) \to b)$

source   Takes a category, returns its source function

target   Takes a category, returns its target function

terminal_obj   The terminal object in a complete category

terminal_arrow   In a complete category, the unique arrow from an object to the terminal object

Appendix D

Answers to Programming Exercises

These are answers to the programming exercises in Chapter 2. They are mainly transcripts of ML sessions.

1.

```
val x = 3; val y = 4 and z = x+1
> val x = 3 : int
> val z = 4 : int
  val y = 4 : int

let val x = 1 and y = 2 in x+y end
> 3 : int

val p = 3 and q = p+1
Type checking error in: p
Unbound value identifier: p

let val (x,y) = (2,3) in 2*x + y end
> 7 : int

let val x = 1 in
   let val y = x+2 in
     let val x = 5 in x+y end end end
> 8 : int
```

```
val (x, y as (_,p)) = ((2,3),(4,(5,6)))
> val p = (5,6) : int * int
  val y = (4,(5,6)) : int * (int * int)
  val x = (2,3) : int * int
```

2.

```
fun sign(n: int) = if n < 0 then false else true

fun absvalue(n:int) = if n < 0 then 0-n else n

fun max(m,n:int) = if m < n then n else m

fun fib(n) = if n=0 then 1 else
                if n=1 then 1 else fib(n-1) + fib(n-2)
```

3.

```
fun numprint(zero) = 0
  | numprint(succ(n)) = 1+ numprint(n)

fun mult(zero,n) = zero
  | mult(succ(m),n) = add(n,mult(m,n))
```

4.

```
fun apply(f)(x) = f(x)
> ('a -> 'b) * 'a -> 'b
fun compose(g,f) = fn x => g(f(x))
> ('b -> 'c) * ('a -> 'b) -> ('a -> 'c)
```

5.

```
fun listmax(nil) = 0 | listmax(x::s) = max(x,listmax(s))
fun sum(nil) = 0 | sum(x::s) = x + sum(s)
fun poly(nil)(x:int) = 0
  | poly(a::s)(x) = a + x * poly(s)(x)
fun reverse(nil) = nil
  | reverse(x::s) = append(reverse(s),[x])
fun maplist(f)(nil) = nil
  | maplist(f)(x::s) = f(x)::maplist(f)(s)
```

The `accumulate` function can be written in two ways depending on which end of the list we start at:

```
fun accumulate(f)(initial)(nil) = initial
  | accumulate(f)(initial)(x::s) =
        f(x,accumulate(f)(initial)(s))
fun accumulate(f)(initial)(nil) = i
  | accumulate(f)(initial)(x::s) =
        accumulate(f)(f(x,initial))(s)
```

6.

```
fun breadth(tip n) = 1
  | breadth(node(s,t)) = breadth(s) + breadth(t)
fun depth(tip n) = 1
  | depth(node(s,t)) = 1 + max(depth(s),depth(t))
fun flatten(tip n) = [n]
  | flatten(node(s,t)) = append(flatten(s),flatten(t))
```

7.

```
abstype Rat = fraction of (int * int)
  with fun intrat(n) = fraction(n,1)
       val zero = intrat(0)
       val one  = intrat(1)
       fun numerator(fraction(n,d)) = n
       fun denominator(fraction(n,d)) = d
       fun add(fraction(n,d),fraction(n',d')) =
                          fraction(n*d'+n'*d,d*d')
       fun minus(fraction(n,d),fraction(n',d')) =
                          fraction(n*d'-n'*d,d*d')
       fun times(fraction(n,d),fraction(n',d')) =
                          fraction(n*n',d*d')
       fun mkrat(n,d) = fraction(n,d)
  end
```

Other functions may be defined in the abstract type. The type could be modified so that fractions are kept in their reduced form or displayed without any common factors between numerator and denominator.

234   *ANSWERS TO PROGRAMMING EXERCISES*

8.

```
fun delete(n)(x,nil) = nil
  | delete(n)(x,a::s) = if n=1
          then if x=a then s else a::delete(n)(x,s)
          else if x=a then a::delete(n-1)(x,s)
                      else a::delete(n)(x,s)
fun sublist(nil,t) = true
  | sublist(a::s,nil) = false
  | sublist(a::s,b::t) =
      if a=b then sublist(s,t) else sublist(a::s,t)
fun number_of_sublists(nil,t) = 1
  | number_of_sublists(a::s,nil) = 0
  | number_of_sublists(a::s,b::t) =
      if a=b then number_of_sublists(s,t) +
                      number_of_sublists(a::s,t)
             else number_of_sublists(a::s,t)
```

9.

```
fun disjoint_union(s,t) =
  union(image(fn x => (x,0))(s),image(fn x => (x,1))(t)

fun product(s,t) = if is_empty(s) then emptyset else
        let val (x,s') = singleton_split(s) in
      union(image(fn y => (x,y))(t),product(s',t)) end

fun powerset(s) =
  if is_empty(s) then singleton(emptyset) else
    let val (x,s') = singleton_split(s)
        val ps' = powerset(s') in
    union(image(fn t => union(singleton(x),t))(ps'),ps')
    end
```

10.

```
datatype BTree =
    empty | tip of int | node of BTree*int*BTree
fun insert(n,empty) = tip(n)
  | insert(n,tip(m)) =
      if n < m then node(tip(n),m,empty)
```

```
                    else node(empty,m,tip(n))
   | insert(n,node(s,m,t)) =
       if n < m then node(insert(n,s),m,t)
                   else node(s,m,insert(n,t))
 fun flatten(empty) = nil | flatten(tip(n)) = [n]
   | flatten(node(s,m,t)) =
               append(flatten(s),m::flatten(t))
 fun maplist(f)(nil) = nil
   | maplist(f)(a::s) = f(a)::maplist(f)(s)
 fun sort(s) = flatten(accumulate(insert)(empty)(s))
```

11.

We give the definition of the map $f \mapsto f^\#$ and then some example uses.

```
fun list_extend(e,star)(f) =
    fn nil => e
    | a::s => star(f(a),list_extend(e,star)(f)(s))

> val list_extend = fn :
    ('a * (('b * 'a) -> 'a)) ->
        (('c -> 'b) -> (('c list) -> 'a))
val length =
  list_extend(0,fn (x,y:int) => x+y)(fn x => 1)
val set_of_elements =
  list_extend(emptyset,fn (x,y) => union(x,y))
            (fn x => singleton(x))
val sum = list_extend(0,fn (x,y:int) => x+y)(fn x=>x)
```

References and Bibliography

- *LNCS = Lecture Notes in Comp. Sci.*, Springer-Verlag, Berlin.

- *LNM = Lecture Notes in Math.*, Springer-Verlag, Berlin.

Abelson, H. and Sussman, G.J, with Sussman, J. (1985) *Structure and Interpretation of Computer Programs.* MIT Press, Cambridge, Mass.

Aczel, P. (1977) The Strength of Martin-Löf's Type Theory with One Universe. *Proc. Symp. Math. Logic, Oulu*, Dept. Philosophy, University of Helsinki, pp. 1–32.

Aczel, P. (1980) Frege Structures and Notions of Proposition. In (eds.) Barwise, Keisler and Kunen, *The Kleene Symposium.* North-Holland, Amsterdam, pp 31–59.

Adámek, J. (1977) Colimits of Algebras Revisited. Bull. *Austral. Math. Soc.*, 17. pp. 433–450.

Adámek, J. (1978) Construction of Free Ordered Algebras. Internal report, Faculty of Elect. Engin. Fel Cvut, Suchbatrarova 2. 166 27 Praha 6, Czechoslovakia.

Adámek, J. and Koubek, V. (1980) Are Colimits of Algebras Simple to Construct? *J. Algebra*, 66, pp. 226–250.

Adámek, J. and Trnkova, V. (1978) Varietors and Machines. Comp. and Inform. Science Tech. Report 78–6, University of Massachusetts at Amherst.

Aho, A.V., Hopcroft, J.E. and Ullman, J.D. (1974) *The Design and Analysis of Computer Algorithms.* Addison-Wesley.

Alagić, S. (1975) Natural State Transformations. *J. Comp. Sys. Sci.*, 10, pp. 266–307.

Arbib, M. and Manes, E. (1974) Machines in a Category: An Expository Introduction. *SIAM Rev.*, 16, 2, pp. 163–192.

Arbib, M. and Manes, E. (1975) *Arrows, Structures and Functors: The Categorical Imperative.* Academic Press, London.

Arbib, M. and Manes, E. (1975a) Adjoint Machines, State-Behavior Machines and Duality. *J. Pure and Applied Algebra*, 6, pp. 313–344.

Backus, J. (1978) Can programming be liberated from the Von Neumann style? *Comm. ACM*, 21, 8, pp. 613–641.

Barr, M. (1970) Coequalisers and Free Triples. *Maths Z.*, 2, 116, pp. 307–332.

Barr, M. and Wells, C. (1985) *Toposes, Triples and Theories.* Grundlehren der mathematischen Wissenschaften, 273, Springer-Verlag, New York.

Barringer, H., Cheng, J.H. and Jones, C.B. (1984) A Logic Covering Undefinedness in Program Proofs. *Acta Informatica*, 21, pp. 251–269.

Barwise, J. (1974) Axioms for Abstract Model Theory. *Annal Math. Logic*, 7, pp. 221–265.

Beeson, M.J. (1985) *Foundations of Constructive Mathematics: Metamathematical Studies.* Springer-Verlag, Berlin.

Beeson, M.J. (1986) Proving Programs and Programming Proofs. *Proc. Logic, Methodology and Philosophy of Science* VII, Elsevier Science. pp. 51–82.

Bénabou, J. (1985) Fibered Categories and the Foundations of Naive Category Theory. *J. Symbolic Logic*, 50, 1, pp. 10–37.

Benecke, K. and Reichel, H. (1982) Equational Partiality. *Algebra Universalis*, 15, pp. 306–358.

Bergstra, J.A. and Klop, J.W. (1982) Conditional Rewrite Rules: Confluency and Termination. Internal report, Mathematisch Centrum, Amsterdam.

Birkhoff, G. (1938) Structure of Abstract Algebras. *Proc. Camb. Phil. Soc.*, 31, pp. 433–454.

Blass, A. (1984) The Interaction between Category Theory and Set Theory. *Proc. of the Special Session on the Mathematical Applications of Category Theory*, 89th Annual Meeting of the American Mathematical Society, (ed.) J.W. Gray, *Contemporary Mathematics*, 30, American Mathematical Society.

Boehm, H., Demers, A. and Donahue, J. (1980) An Informal Description of Russell. Technical report, Dept. Computer Science, Cornell University.

de Bruijn, N.G. (1980) A Survey of the Project Automath. In (eds.) J.R. Hindley and J.P. Seldin, *To H.B. Curry: Essays on Combinatory Logic, Lambda Calculus and Formalisms*, Academic Press, London.

Burmeister, P. (1982) Partial Algebras – A Survey of a Unifying Approach towards a Two-valued Model Theory for Partial Algebras. *Algebra Universalis*, 15, pp. 306–358.

Burmeister, P. (1986) *A Model Theoretic Oriented Approach to Partial Algebras*. Akademie-Verlag, Berlin.

Burstall, R.M. (1980) Electronic Category Theory. Invited paper, *Proc. 9th Annual Symposium on the Mathematical Foundations of Computer Science, Rydzyua, Poland*.

Burstall, R.M. and Darlington, J. (1977) A Transformation System for Developing Recursive Programs. J. ACM, 24, pp. 44–67.

Burstall, R.M. and Goguen, J.A. (1977) Putting Theories Together to Make Specifications. *Proc. 5th International Joint Conference on Artificial Intelligence, Boston, Mass.*, pp. 1045–1058.

Burstall, R.M. and Goguen, J.A. (1980) The Semantics of Clear, A Specification Language. *Proc. Advanced Course on Abstract Software Specifications, LNCS* 86.

Burstall, R.M. and Goguen, J.A. (1981) An Informal Introduction to Specifications using Clear. In (eds.) R.S. Boyer and J.S. Moore, *The Correctness Problem in Computer Science*, Academic Press, London, pp. 185–215.

Burstall, R.M. and Lampson, B. (1984) A Kernel Language for Abstract Data Types and Modules. *Proc. of Semantics of Data Types Conf., Sophia-Antipolis, France, LNCS* 173.

Burstall, R.M. and Landin, P.J. (1969) Programs and Their Proofs: An Algebraic Approach. In (eds.) B. Meltzer and D. Michie, *Machine Intelligence* 4, Edinburgh University Press, pp. 17–44.

Burstall, R.M., MacQueen, D.B. and Sannella, D.T. (1980) Hope: An Experimental Applicative Language. *Proc. ACM Lisp Conf., Stanford University, California.*

Burstall, R.M. and Thatcher, J.W. (1974) The Algebraic Theory of Recursive Program Schemes. In *Category Theory Applied to Computation and Control. LNCS* 25.

Cardelli, L. and Wegner, P. (1985) On Understanding Types, Data Abstraction, and Polymorphism. *ACM Computing Surveys*, 17, 4.

Cartmell, J. (1978) Generalised Algebraic Theories and Contextual Categories. DPhil thesis, Oxford University.

Cartmell, J. (1985) Reduction Rules for Cartesian Closed Categories. Preprint, Dept. Comp. Sci., Edinburgh University.

Church, C. (1941) The Calculi of Lambda-Conversion. *Ann. Math. Studies*, Princeton University Press.

Codd, E.F. (1970) A Relational Model of Data for Large Shared Data Banks. *Comm. ACM*, 13, 6, pp. 377–387.

Cohn, P.M. (i965) *Universal Algebra.* Harper & Row, New York. Republished 1981 by D. Reidel, Dordrecht, Holland.

Colmerauer, A. et al. (1973) Etude et Realisation d'un Système PROLOG. *Convention de Research IRIA-Sesori* 77030.

Constable, R.L. and Bates, J.L. (1984) The Nearly Ultimate PRL. Technical report, 83–551, Cornell University.

Constable, R.L. et al. (1985) *Implementing Mathematics with the Nuprl Proof Development System.* Prentice Hall, Englewood Cliffs, NJ.

Coquand, T. (1986) An Analysis of Girard's Paradox. *Proc. 1st Logic in Computer Science Symposium. Boston, Mass.*, IEEE, Ithaca, NY.

Coquand, T. and Huet, G. (1985) Constructions: A Higher Order Proof System for Mechanising Mathematics. *EUROCAL85, Linz, LNCS* 203.

Curien, P.-L. (1986) Categorical Combinators, Sequential Algorithms and Functional Programming. *Research Notes in Theoretical Computer Science.* Pitman, London.

Curry, H.B. and Feys, R. (1968) *Combinatory Logic.* 1, North-Holland, Amsterdam.

Darlington, J., Henderson, P. and Turner, D.A. (1982) *Functional Programming and its Applications: An Advanced Course.* CUP.

Dewar, R.B.K. (1978) The SETL Programming Language. Technical report, Courant Inst. of Mathematical Sciences, New York Univ.

Dubuc, E.J. (1974) Free Monoids. *J. Algebra* 29, pp. 208–228.

Dybjer, P (1985) Program Verification in a Logical Theory of Constructions. In *Functional Programming Languages and Computer Architectures LNCS* 201.

Dybjer, P (1986) Category Theory and Programming Language Semantics: An Overview. *Proc. Summer Workshop on Category Theory and Computer Programming 1985, LNCS* 240. pp. 165–181.

Dyckhoff, R. (1985) Category theory as an extension of Martin-Löf type theory. Internal report CS 85/3, Dept. of Computational Science, University of St. Andrews, Scotland.

Ehrig, H., Pfender, M. and Schneider, H. (1973) Graph Grammars: An Algebraic Approach. *Proc. 14th. Annual IEEE Symposium on Switching and Automata Theory*, pp. 167-180.

Eilenburg, S. and Mac Lane, S. (1945) A General Theory of Natural Equivalences. *Trans. Am. Math. Soc.*, 58, pp. 231–294.

Eilenberg, S. and Moore, J.C. (1965) Adjoint Functors and Triples. *Illinois J. Math.*, 9, pp. 381–398.

Elgot, C.C. (1971) Algebraic Theories and Program Schemes. In (ed.) Engeler, *Symp. on the Semantics of Algorithmic Languages, LNM* 188, pp. 71–88.

Eriksson, L.H. (1984) Synthesis of a Unification Algorithm in a Logic Programming Calculus. *J. Logic Programming*, 1, 1.

Ershov, Y.L. (1973) Theorie der Numerierungen I. *Z. Mathematische Logik* 19.

Ershov, Y.L. (1975) Theorie der Numerierungen II. *Z. Mathematische Logik* 21.

Feferman, S. (1969) Set-Theoretical Foundations of Category Theory. In (ed.) S. Mac Lane, *Reports of the Mid-West Category Seminar III, LNM* 106, pp 201-247.

Feferman, S. (1975) A Language and Axioms for Explicit Mathematics. In *Algebra and Logic, LNM* 450. pp. 87–139.

Feferman, S. (1979) Constructive Theories of Functions and Classes. In (eds.) M. Boffa, D. van Dalen and K. McAloon, *Logic Colloquium '80*, pp. 95–128. North-Holland, Amsterdam.

Freyd, P. (1964) *Abelian Categories: An Introduction to the Theory of Functors*. Harper & Row, New York.

Freyd, P. (1972) Aspects of Topoi. *Bull. Aust. Math. Soc.*, 7, pp. 1–76.

Futatsugi, K., Goguen, J.A., Jouannaud, J.-P. and Meseguer, J. (1985) Principles of OBJ2. In *Proc. Symposium on Principles of Programming Languages, ACM*, pp. 52–66.

Girard. J.Y. (1972) Interprétation fonctionelle et élimination des coupures dans l'arithmétique d'ordre supérieur. PhD thesis, Paris.

Glaser, H., Hankin, C. and Till, D. (1984) *Principles of Functional Programming*. Prentice Hall International, Hemel Hempstead.

Goguen, J.A. (1971) Mathematical Representation of Hierarchically Organized Systems. In (ed.) E. Atlinger, *Global Systems Dynamics*, pp. 112–128. S. Karger.

Goguen, J.A. (1973) Realization is Universal. *Math. Sys. Th.*, 6, pp. 359–374.

Goguen, J.A. (1978) Order Sorted Algebra. UCLA Computer Science Dept., Semantics and Theory of Computation report 14; to appear in *J. Comp. Sys. Sci.*

Goguen, J.A. (1988) Higher Order Functions Considered Unnecessary for Higher Order Programming. Internal report SRI-CSL-88-1, SRI International.

Goguen, J.A. and Burstall, R.M. (1983) Introducing Institutions. *Proc. Logics of Programming Workshop, Carnegie-Mellon University.*

Goguen, J.A. and Burstall, R.M. (1984) Some Fundamental Tools for the Semantics of Computation, Part 1: Comma Categories, Colimits, Signatures and Theories. *Th. Comp. Sci.*, 31, pp. 175–209.

Goguen, J.A. and Ginali, S. (1978) A Categorical Approach to General Systems Theory. In (ed.) G. Klir, *Applied General Systems Research* pp. 257–270, Plenum, New York.

Goguen, J.A. and Meseguer, J. (1985) Completeness of Many-sorted Equational Logic. *Houston J. Math.*, 11, 3, pp. 307-334.

Goguen, J.A. and Meseguer, J. (1988) Semantics of Computation. Preprint, book in preparation.

Goguen, J.A. and Tardo, J. (1979) An Introduction to OBJ: A Language for Writing and Testing Software. In *Specification of Reliable Software*, pp. 170–189, IEEE, Ithaca, NY.

Goguen, J.A., Thatcher, J.W., Wagner, E.G. and Wright, J.B. (1977) Initial Algebra Semantics and Continuous Algebras. *J. ACM* 24, 1, pp. 68–95.

Goguen, J.A., Thatcher, J.W. and Wagner, E.G. (1978) An Initial Algebra Approach to the Specification, Correctness and Implementation of Abstract Data Types. In (ed.) R. Yeh, *Current Trends in Programming Methodology* Prentice Hall, Englewood Cliffs, NJ. pp. 80–149.

Goldblatt, R. (1979) *Topoi – The Categorial Analysis of Logic.* Studies in Logic and the Foundations of Mathematics, 98, North-Holland, Amsterdam.

Gordon, M.J.C. (1979) *The Denotational Description of Programming Languages.* Springer-Verlag, Berlin.

Gordon, M.J.C., Milner, R. and Wadsworth, C.P. (1979) *Edinburgh LCF. LNCS* 78.

Grothendieck, A. (1963) Catégories fibrées et descente. Revêtements étales et group fondamental. *Séminaire de Gémétrie Algébrique du Bois-Marie 1960/61* (SGA 1), exposé VI, 3rd ed., Institut des Hautes Études Scientifiques, Paris; reprinted in *LNM* 224 (1971).

Hagino, T. (1987) Categorical Data Types. PhD thesis, Internal report, LFCS, Dept. Comp. Sci., University of Edinburgh.

Hagino, T. (1987a) A Typed Lambda Calculus with Type Constructors. *Proc. Summer Conference on Category Theory and Computer Science, Edinburgh, LNCS* 283.

Hamza, T.T.A. (1985) Normalisation Techniques in Proof Theory and Category Theory – An Implementation and Applications. PhD thesis, University of St. Andrews, Scotland.

Harper, R., Honsell, F. and Plotkin, P. (1987) A Framework for Defining Logics. *Proc. Symposium on Logic in Computer Science, June 22-25*, IEEE, Ithaca, NY.

Harper, R. and Mitchell, K. (1986) Introduction to Standard ML. Unpublished Report, University of Edinburgh.

Henderson, P. (1980) *Functional Programming: Applications and Implementation.* Prentice Hall International, Hemel Hempstead.

Herbrand, J. (1930) Recherches sur la Théorie de la Démonstration. thesis, University of Paris, In (1968) *Ecrits logique de Jacques Herbrand*, PUF, Paris.

Herrlich, H. and Strecker, G.E. (1973) *Category Theory.* Allyn and Bacon.

Hindley, R. (1969) The Principal Type-scheme of an Object in Combinatory Logic. *Trans. Am. Math. Soc.*, 146, pp. 29–40.

Hintze, S. (1981) Kategorientheoretische Beschreibung rekursiver Funktionale. Unpublished paper.

Hook, J.G. (1984) Understanding Russell – A First Attempt. *Proc. Semantics of Data Types Conf., Sophia-Antipolis, France, LNCS* 173.

Hopcroft, J.E. and Ullman, J.D. (1979) *Introduction to Automata Theory, Languages, and Computation.* Addison-Wesley.

Huet, G. (1975) A Unification Algorithm for Typed Lambda Calculus. *Th. Comp. Sci.*, 1, 1, pp. 27–57.

Huet, G. (1976) Résolution d'Equations dans les Languages d'Ordre $1, 2, \ldots, \omega$. thesis, Specialité Maths., University of Paris, VII.

Huet, G. (1980) Confluent Reductions: Abstract Properties and Applications to Term Rewriting Systems. *J. ACM*, 27, 4, pp. 797–821.

Huet, G. (1986) Formal Structures for Computation and Deduction. Working Paper for International Summer School on Logic of Programming and Calculi of Discrete Design, Marktoberdorf, Germany.

Huet, G. and Oppen, D.C (1980) Equations and Rewrite Rules: A Survey. In (ed.) R. Book, *Formal Languages: Perspectives and Open Problems*, Academic Press.

Hyland, J.M.E. (1982) The effective topos. In (eds.) A.S. Troelstra and D. van Dalen, *The L.E.J. Brouwer Centenary Symposium* North-Holland, Amsterdam, pp. 165–216.

Isbell, J. (1957) Some remarks concerning categories and subspaces. *Can. J. Math.*, 9, pp. 563–577.

Johnstone, P.T. (1977) *Topos Theory.* Academic Press, London.

Jones, C.B. (1986) *Systematic Software Development using VDM.* Prentice Hall, Hemel Hempstead.

Joyal, A. (1981) Une Theorie Combinatoire des Séries Formelles. *Adv. Math.*, 42, pp. 1–82.

Kan, D.M. (1958) Adjoint Functors. *Trans. Am. Math. Soc.*, 87, pp. 294–329.

246 REFERENCES AND BIBLIOGRAPHY

Kapur, D., Musser, D.R. and Stepanov, A.A. (1981) Operators and Algebraic Structures. *Proc. ACM Conf. Functional Prog. Languages.*

Kleisli, H. (1965) Every Standard Construction is Induced by a Pair of Adjoint Functors. *Proc. Am. Math. Soc.*, 16, pp. 544–546.

Knuth, D.E. (1973) *Searching and Sorting.* Addison-Wesley.

Knuth, D.E. and Bendix, P.B. (1970) Simple Word Problems in Universal Algebras. In (ed.) J. Leech *Computational Problems in Abstract Algebra*, Pergamon Press, Oxford, pp. 263–297.

Kruse, R.L. (1987) *Data Structures and Program Design.* Prentice Hall, Englewood Cliffs, NJ.

Lambek, J. (1969) Deductive systems and categories II. *LNM* 86, pp 76–122.

Lambek, J. (1980) From λ-calculus to Cartesian Closed Categories. In (eds.) J.R. Hindley and J.P. Seldin, *To H.B. Curry: Essays on Combinatory Logic, Lambda-Calculus and Formalism.* Academic Press, pp. 375–402.

Lambek, J. and Scott, P.J. (1986) *Introduction to Higher Order Categorical Logic.* CUP.

Landin, P.J. (1966) The Next 700 Programming Languages. *Comm. ACM*, 9, 3, pp. 157–166.

Lawvere, F.W. (1963) Functorial Semantics of Algebraic Theories. PhD thesis, Columbia University.

Lawvere, F.W. (1963a) Functorial Semantics of Algebraic Theories. *Proc. Nat. Acad. Sci.*, 50, pp. 869–872.

Lawvere, F.W. (1966) The Category of Categories as a Foundation for Mathematics. In (eds) S. Eilenberg et al., *Proc. Conf. on Categorical Algebra, La Jolla, 1965.* Springer-Verlag.

Lawvere, F.W. (1970) Equality in Hyperdoctrines and the Comprehension Schema as an Adjoint Functor. *Proc. Symp. in Pure Math., XVII: Applications of Categorical Algebra*, Am. Math. Soc., pp. 1–14.

Lawvere, F.W. (1974) Variable Quantities and Variable Structures in Topoi. In *Algebra, Topology, and Category Theory*, Academic Press, pp. 101–132.

Lehmann, D.J. (1977) Algebraic Structures for Transitive Closure. *Th. Comp. Sci.*, 4, 1, pp. 59–76.

Levi, G. and Sirovich, F. (1975) Proving Program Properties, Symbolic Evaluation and Logical Procedural Semantics. *Proc. Math. Found. Comp. Sci., LNCS* 32.

McCarthy, J. (1960) Recursive Functions of Symbolic Expressions and Their Computation by Machine. *Comm. ACM*, 3, 4, pp. 184–195.

Mac Lane, S. (1948) Groups, Categories and Duality. *Proc. Nat. Acad. Sci.*, 34, pp. 263–267.

Mac Lane, S. (1950) Duality for Groups. *Bull. Am. Math. Soc.*, 56, pp. 485–516.

Mac Lane, S. (1965) Categorical Algebra. *Bull. Am. Math. Soc.*, 71, pp. 40–106.

Mac Lane, S. (1971) *Categories for the Working Mathematician.* Springer-Verlag, New York.

Mac Lane, S. (1975) Sets, Topoi and Internal Logic in Categories. *Proc. Logic Coll., Bristol, 1973*, North-Holland, Amsterdam.

McLarty, C. (1986) Left Exact Logic. *J. Pure and Appl. Algebra*, 41, pp. 63–66.

MacQueen, D. (1985) Modules for Standard ML. *Polymorphism*, 2, 2. Earlier version in *Proc. ACM Symp. on Lisp and Functional Programming, 1984, Austin, Texas.*

Manes, E.G. (1976) *Algebraic Theories.* Springer-Verlag, New York.

Manna, Z. and Waldinger, R. (1980) Deductive Synthesis of the Unification Algorithm. SRI research report.

Martelli, A. and Montanari, U. (1982) An Efficient Unification Algorithm. *ACM Trans. Prog. Lang. Sys.*, 4, 2, pp. 258–282.

Martin-Löf, P. (1975) An Intuitionistic Theory of Types: Predicative Part. In (eds.) H.E. Rose and J.C. Shepherdson, *Logic Colloquium, Bristol, 1973*, North-Holland, Amsterdam, pp. 73–118.

Martin-Löf, P. (1982) Constructive Mathematics and Computer Science. *6th Int. Congress for Logic, Methodology and Philosophy of Science, Hannover, 1979*. North-Holland, Amsterdam.

Martin-Löf, P. (1984) *Intuitionistic Type Theory*, Bibliopolis, Naples.

Milner, R. (1978) A Theory of Type Polymorphism in Programming. *J. Comp. Sys. Sci.*, 17, 3, pp. 348–375.

Milner, R. (1984) A Proposal for Standard ML. *Proc. ACM Symp. on Lisp and Functional Programming, Austin, Texas.*

Moggi, E. (1985) Partial Morphisms in Categories of Effective Objects. Unpublished paper, Dept. Comp. Sci., University of Edinburgh.

Mulry, P. (1981) Generalized Banach-Mazur Functionals in the Topos of Recursive Sets, *J. Pure and Appl. Algebra*, 26.

Oles, F.J. (1985) Type Algebras, Functor Categories and Block Structure. In (eds.) M. Nivat and J.C. Reynolds, *Algebraic Methods in Semantics*, CUP, Chapter 15.

Paterson, M.S. and Wegman, M.N. (1978) Linear Unification. *J. Comp. Sys. Sci.*, 16, 2, pp. 158–167.

Paulson, L.C. (1985) Verifying the Unification Algorithm in LCF. *Sci. Comp. Programming*, 5, 2, pp. 143–170.

Petersson, K. (1982,84) A Programming System for Type Theory. LPM memo 21, Dept. Comp. Sci., Chalmers Institute of Technology, Göteborg.

Pitt, A.M. (1987) Polymorphism is Set Theoretic, Constructively. *Proc. Summer Conference on Category Theory and Computer Science, Edinburgh 1987, LNCS* 283.

Plotkin, G. (1969) A Note on Inductive Generalisation. In (eds.) B. Meltzer and D. Michie, *Machine Intelligence* 5, Edinburgh University Press, pp. 153–163.

Plotkin, G. (1985) Denotational Semantics with Partial Functions. *Lectures at the CSLI Summer School, Stanford, July 1985.*

Poigné, A. (1984) On Specifications, Theories and Models with Higher Types. Preprint, Imperial College, London.

Reichel, H. (1980) Initially-Restricting Algebraic Theories. Unpublished report, TH. 'Otto von Guericke', Sektion Math/Physik, Magdeburg, DDR.

Reichel, H. (1987) *Initial Computability, Algebraic Specifications and Partial Algebras.* Clarendon Press, Oxford.

Reynolds, J.C. (1974) Towards a Theory of Type Structure. *Colloque sur la programmation, Paris, LNCS* 19, pp 408–425.

Reynolds, J.C. (1981) *The Craft of Programming.* Prentice Hall, Hemel Hempstead.

Reynolds, J.C. (1983) Types, Abstraction and Parametric Polymorphism. *Proc. IFIP Congress, Paris.*

Robinson, E.P. and Rosolini, G. (1986) Categories of Partial Maps. Quaderno del Dipartimento di Matematica, 18, Università di Parma.

Robinson, J.A. (1965) A Machine-oriented Logic Based on the Resolution Principle. *J. ACM*, 12, 1, pp. 23–41.

Robinson, J.A. and Wos, L.T. (1969) Paramodulation and Theorem Proving in First-Order Theories with Equality. In (eds) B. Meltzer and D. Michie, *Machine Intelligence* 4, Edinburgh University Press, pp. 135–150.

Rydeheard, D.E. (1981) Applications of Category Theory to Programming and Program Specification, PhD thesis. Dept. Comp. Sci., University of Edinburgh.

Rydeheard, D.E. and Burstall, R.M. (1985) The Unification of Terms: A Category-theoretic Algorithm. Internal report UMCS-85-8-1, Dept. Comp. Sci., University of Manchester.

Rydeheard, D.E. and Burstall, R.M. (1985) Monads and Theories – A Survey for Computation. In (eds) M. Nivat and J.C. Reynolds, *Algebraic Methods in Semantics*, CUP, Chapter 16.

Rydeheard, D.E. and Burstall, R.M. (1986) A Categorical Unification Algorithm. *Proc. Summer Workshop on Category Theory and Computer Programming, 1985, LNCS* 240, pp. 493–505.

Rydeheard, D.E. and Stell, J.G. (1987) Foundations of Equational Deduction: A Categorical Treatment of Equational Proofs and Unification Algorithms. *Proc. Summer Conference on Category Theory and Computer Science, Edinburgh, 1987, LNCS* 283.

Sannella, D.T. (1982) Semantics, Implementation and Pragmatics of Clear, a Program Specification Language. PhD thesis, Dept. Comp. Sci., University of Edinburgh.

Sannella, D.T. (1984) A Set-theoretic Semantics for Clear. *Acta Informatica*, 21, 5, pp. 443–472.

Sannella, D. and Tarlecki, A. (1986) Extended ML: An Institution-Independent Framework for Formal Program Development. *Proc. Summer Workshop on Category Theory and Computer Science, Surrey, 1985, LNCS* 240.

Sannella, D.T. and Wirsing, M. (1983) A Kernel Language for Algebraic Specification and Implementation. Report CSR-131-83, Dept. Comp. Sci., University of Edinburgh. Extended abstract in: *Proc Int. Conf. on Foundations of Computation Theory, Borgholm, Sweden, LNCS* 158, pp. 413-427.

Schubert, H. (1972) *Categories.* Springer-Verlag, Berlin.

Scott, D.S. (1979) Identity and Existence in Intuitionistic Logic. In *Applications of Sheaves, LNM* 753, pp. 660–696.

Scott, D.S. (1980) Relating Theories of the λ-calculus. In (eds.) J.R. Hindley and J.P. Seldin, *To H.B. Curry: Essays on Combinatory Logic, Lambda-Calculus and Formalism.* Academic Press, London, pp. 403–450.

Seely, R.A.G. (1983) Hyperdoctrines, Natural Deduction and the Beck Condition. *Z. Math. Logik*, 29, pp. 505–542.

Seely, R.A.G. (1984) Locally Cartesian Closed Categories and Type Theory. Math. *Proc. Camb. Phil. Soc.*, 95, pp. 33–48.

Seely, R.A.G. (1987) Modelling Computations: A 2-Categorical Framework. *Proc. Symp. Logic in Computer Science, June 22–25, 1987*, IEEE, Ithaca, NY.

Siekman, J.H. (1984) Universal Unification. *7th Int. Conf. Automated Deduction, LNCS* 170.

Smyth, M.B. and Plotkin, G.D. (1982) The Category-theoretic Solution of Recursive Domain Equations. *SIAM J. Computing*, 11, pp. 761–783.

Strachey, C. (1967) Fundamental Concepts in Programming Languages. Lecture notes, *International Summer School in Computer Programming, Copenhagen.*

Szabo, M. (1978) *Algebra of Proofs.* North-Holland, Amsterdam.

Tarjan, R. (1972) Depth-First Search and Linear Graph Algorithms. *SIAM J. Computing*, 1, pp. 146-160.

Turner, D.A. (1979) A New Implementation Technique for Applicative Languages. *Software Practice and Experience*, 9, 10.

Turner, D.A. (1986) Miranda: A Non-Strict Functional Language with Polymorphic Types. *Proc. STACS 86, LNCS* 210.

Wand, M. (1979) Final Algebra Semantics and Data Type Extensions. *J. Comp. Sys. Sci.*, 19, pp. 27–44.

Warren, D.H.D., Pereira, L.M. and Pereira, F. (1977) PROLOG – The Language and its Implementation Compared with Lisp. *Proc. Symp. on Artificial Intelligence and Prog. Languages, University of Rochester, NY.* Appeared in joint issue: *SIGPLAN Notices* 12, 8, and *SIGART Newsletter*, 64.

Warshall, S. (1962) A Theorem on Boolean Matrices. *J. ACM*, 9, 1, pp. 11–12.

Watjen, D. and Struckmann, W. (1982) An Algorithm for Verifying Equations of Morphisms in a Category. *Inform. Process. Letters*, 14, 3, pp. 104–108.

Wikström, A. (1987) *Functional Programming using Standard ML.* Prentice Hall International, Hemel Hempstead.

Winskel, G. (1986) Category Theory and Models of Parallel Computation. *Proc. Summer Workshop on Category Theory and Computer Programming, Surrey, 1985, LNCS* 240, pp. 266–281.

Winskel, G. (1987) Relating Two Models of Hardware. *Proc. Summer Conference on Category Theory and Computer Science, Edinburgh, 1987, LNCS* 283, pp. 98–113.

Wright, J.B., Goguen, J.A., Thatcher, J.W. and Wagner, E.G. (1976) Rational Algebraic Theories and Fixed Point Solutions. *Proc. 17th IEEE Symp. Foundations of Computer Science, Houston, Texas.*

Yelick, K. (1985) Combining Unification Algorithms for Confined Regular Equational Theories. *Proc. 1st International Conf. Rewriting Techniques and Applications, Dijon, France, LNCS* 202, pp. 365-380.

Index